1,000,000 Books

are available to read at

www.ForgottenBooks.com

Read online
Download PDF
Purchase in print

ISBN 978-0-282-60756-2
PIBN 10858604

This book is a reproduction of an important historical work. Forgotten Books uses state-of-the-art technology to digitally reconstruct the work, preserving the original format whilst repairing imperfections present in the aged copy. In rare cases, an imperfection in the original, such as a blemish or missing page, may be replicated in our edition. We do, however, repair the vast majority of imperfections successfully; any imperfections that remain are intentionally left to preserve the state of such historical works.

Forgotten Books is a registered trademark of FB &c Ltd.
Copyright © 2018 FB &c Ltd.
FB &c Ltd, Dalton House, 60 Windsor Avenue, London, SW19 2RR.
Company number 08720141. Registered in England and Wales.

For support please visit www.forgottenbooks.com

1 MONTH OF FREE READING

at

www.ForgottenBooks.com

By purchasing this book you are eligible for one month membership to ForgottenBooks.com, giving you unlimited access to our entire collection of over 1,000,000 titles via our web site and mobile apps.

To claim your free month visit:
www.forgottenbooks.com/free858604

* Offer is valid for 45 days from date of purchase. Terms and conditions apply.

English
Français
Deutsche
Italiano
Español
Português

www.forgottenbooks.com

Mythology Photography **Fiction**
Fishing Christianity **Art** Cooking
Essays Buddhism Freemasonry
Medicine **Biology** Music **Ancient Egypt** Evolution Carpentry Physics
Dance Geology **Mathematics** Fitness
Shakespeare **Folklore** Yoga Marketing
Confidence Immortality Biographies
Poetry **Psychology** Witchcraft
Electronics Chemistry History **Law**
Accounting **Philosophy** Anthropology
Alchemy Drama Quantum Mechanics
Atheism Sexual Health **Ancient History**
Entrepreneurship Languages Sport
Paleontology Needlework Islam
Metaphysics Investment Archaeology
Parenting Statistics Criminology
Motivational

A Treatise

UPON

ANALYTICAL MECHANICS;

BEING THE FIRST BOOK

OF THE

MECHANIQUE CELESTE

OF

Pierre Simon, Marquis de

P. S. LAPLACE,

Member of the Institute and of the Bureau of Longitude
of France, &c. &c.

TRANSLATED AND ELUCIDATED

WITH

Explanatory Notes,

BY THE REV. JOHN TOPLIS, B. D.

FELLOW OF QUEEN'S COLLEGE, CAMBRIDGE.

Nottingham;

PRINTED BY H. BARNETT.

SOLD BY LONGMAN, HURST, REES, ORME, AND BROWN,
AND CRADDOCK AND JOY, LONDON; AND
J. DEIGHTON, CAMBRIDGE.

1814

Phys 778.14

PREFACE.

IT has been for some time a subject of complaint amongst mathematical readers, that, although the analytical sciences have been investigated with the greatest ardour and success for a length of time by men of the most eminent talents upon the continent, yet scarcely any works exist in the English language in which the improvements made by them are noticed.

As their notation and peculiar modes of proceeding are different from those used by English Mathematicians; I conceive that a translation of an elementary treatise upon analytical mechanics by one of the most distinguished of the continental Analysts, with notes that shall enable the reader to understand it with greater facility, will render an acceptable service to those who are desirous of being in some degree acquainted with their merits.

It is hoped that when this work is understood, the obstacles which may be met with in reading the Méchanique Céleste, or the Méchanique Analytique, will principally arise from the difficulty of integrating the equations of which the authors make use. This can only be obviated by a diligent perusal of some of the treatises upon the integral calculus, given by Euler, Waring, Cousin, Legendre, and Lacroix; particularly the Traité du Calcul Différentiel et Intégral by the last mentioned writer.

To say any thing relative to the merits of such productions as the Méchanique Céleste or the Méchanique Analytique would be superfluous. They are so great as to induce an eminent Analyst* to assert, that all elementary treatises upon mechanics should be composed as preparatory to a perusal of them.

The work which is here translated will serve to give those readers who are unacquainted with the writings of the continental Analysts, some idea of the elegance and extent of which their methods are capable. I have only to regret that the task of presenting it has not been attempted by abler hands; but, from some excellent analytical specimens already before the public, little

---

* It is now certain that the Mechanique Analytique and the Mechanique Celeste are the true sources from which a complete and methodical knowledge of all the properties of the equilibrium and of the motion of bodies either solid or fluid, which objects form the principal application of transcendental analysis, can be obtained; it is therefore necessary that in future the elementary treatises should be composed with the view of leading to these works.

Discours preliminaire to the Traite elementaire de Calcul Differentiel et de Calcul Integral par S. F. Lacroix, Edit. 1802.

doubt can be entertained of it soon being followed by original treatises upon the same subject, by writers well qualified for the undertaking.—From the rapidly increasing taste for the works of the continental Mathematicians, there is every reason to hope, that the time is not far distant when the analytical sciences will again flourish in the country of their illustrious founder.

With respect to the notes, the whole of which I have added to the treatise of Laplace, it may be proper to observe, that they are intended in some degree to facilitate the reading of the text to those students whose information is not supposed to extend beyond the elementary principles of mechanics and of fluxions as taught in this island.—They contain some additions to the original work. These are particularly necessary, as the treatise of Laplace was merely intended for an introduction to the Méchanique Céleste In making them I have derived considerable assistance from the works of the most eminent writers upon mechanics, particularly Lagrange.

As these notes were merely composed with the hope of promoting a favourite science, of which every one who is acquainted with the confined sale of the generality of mathematical works must be well aware, I hope the reader will receive with indulgence any errors into which I may unintentionally have fallen; particularly when he is informed, that they were written at intervals under very unfavourable circumstances arising from the care of a school and other duties, which the " res angusta domi" rendered unavoidable.

Nottingham Grammar School.
May 28th. 1814.

THE
MECHANICS
OF
LAPLACE.

CHAP. I.

Of the equilibrium and of the composition of the forces which act upon a material point.

1. A Body appears to us to be in motion when it changes its situation relative to a system of bodies which we suppose to be at rest: but as all bodies, even those which seem to be in a state of the most absolute rest, may be in motion; we conceive a space, boundless, immoveable, and penetrable to matter: it is to the parts of this real or ideal space that we by imagination refer the situation of bodies; and we conceive them to be in motion when they answer successively to different parts of space.

The nature of that singular modification in consequence of which bodies are transported from one place to another, is, and always will be unknown: we have designated it by the name of force; and we are not able to determine any thing more than its effects, and the laws of its action. The effect of a force acting upon a material point is, if no obstacle opposes, to put it into motion; the direction of the force is the right line which it tends to make the point describe. It is evident that if two forces act in the same direction, their effect is the sum of the two forces, and that if they act in a contrary direction, the point is moved by a force represented by their difference. If their directions form an angle with each other, a force results the direction of which is a mean between the directions of the composing forces. Let us see what is this resultant and its direction.

For this purpose, let us consider two forces x and y acting at the same time upon a material point M, and forming a right angle with each other. Let z represent their resultant, and θ the angle which it makes with the direction of the force x; the two forces x and y being given, the angle θ will be determined, as well as the resultant z; in short there exists amongst the three quantities x, z, and θ a relation which it is required to know.

Let us then suppose the forces x and y infinitely small, and equal to the differentials dx and dy; let us suppose again that x becoming successively dx, $2dx$, $3dx$, &c. y becomes dy, $2dy$, $3dy$, &c.; it is evident that the angle θ will be always the same, and that the resultant z will become successively dz, $2dz$, $3dz$, &c.; therefore in the successive increments of the three forces x, y, and z, the ratio of x to z will be constant, and can be expressed by a function * of θ which we will re-

---

* Every expression in which any number of indeterminate quantities enter in any manner, is called a function of the indeterminate quantities. Thus x^3, a^x, $a+bx$, sin. x and

present by $\varphi(\theta)$; we shall therefore have $x = z\,\varphi(\theta)$; an equation in which x may be changed into y, provided that at the same time we change the angle θ into $\frac{\pi}{2} - \theta$, π being the semi-circumference of a circle whose radius is unity.

Moreover the force x may be considered as the resultant of two forces x' and x'', of which the first x' is directed along the resultant z, and the second x'' perpendicular to it. The force x which results from these two new forces forming the angle θ with the force x' and the angle $\frac{\pi}{2} - \theta$ with the force x'', we shall have

$$x' = x \cdot \varphi(\theta) = \frac{x^2}{z}\,;\quad x'' = x \cdot \varphi\left(\frac{\pi}{2} - \theta\right) = \frac{xy}{z}\,;$$

these two forces may be substituted for the force x.

In like manner two new forces y' and y'' may be substituted for y, the first being equal to $\frac{y^2}{z}$ and directed along z, and the second equal to $\frac{xy}{z}$ and perpendicular

---

logarithm of $(a+bx)$ are called functions of x; and $ax+y$, $(x+y)^3$, sin. $(ax+y)$ and log. $(ax+y^2)$ are called functions of x and y. One quantity is called an explicit function of another quantity or quantities, when we directly perceive how it is formed from the other quantity or quantities. Thus in the expressions $y = ax^2 + bx + c$ and $y = axz + bx^2 + cz^2$; we have first y an explicit function of x, and next y an explicit function of x and z. When we do not directly see how one quantity is formed from others, but must find it by an algebraical process, we call that quantity an implicit function of the others. Thus in the first of the foregoing equations, x is an implicit function of y, and in the second, an implicit function of y and z.

to z; we shall therefore have, instead of the two forces x and y, the four following:

$$\frac{x^2}{z}, \frac{y^2}{z}, \frac{xy}{z}, \frac{xy}{z};$$

the two last acting in contrary directions destroy each other; the two first acting in the same direction, when added together form the resultant z; we have therefore

$$x^2 + y^2 = z^2;$$

from which it follows that the resultant of the two forces x and y is represented in quantity by the diagonal of a rectangle whose sides represent these forces.

Let us proceed to determine the angle θ.* If we increase the force x by its differential dx, without altering the force y, this angle will be diminished by the indefinitely small quantity $d\theta$, but it is possible to suppose the force dx resolved into two, one dx' in the direction of z, and the other dx'' perpendicular to it; the point M will then be acted upon by the two forces $z+dx'$ and dx'' which are perpendicular to each other, and the resultant of these two forces which we represent by z' will make with dx'' the angle $\frac{\pi}{2}-d\theta$; we shall have therefore by what precedes

$$dx'' = z' \cdot \varphi\left(\frac{\pi}{2} - d\theta\right)$$

the function $\varphi\left(\frac{\pi}{2} - d\theta\right)$ is consequently indefinitely

---

* That the reasoning in the proof of the direction of the resulting force may be more readily comprehended, I have given a diagram, (*fig.* 1.) in which we may suppose Mx or $bz = x$, Mb or xz or $ac = y$, $Mz = z$, xa or $zc = dx$, $zz' = dx'$, $z'c = dx''$, $Mc =$ resultant of $z + dx'$ and dx'', angle $zMx = \theta$, and angle $zMc = -d\theta$.

small, and of the form $-kd\theta$, k being a constant quantity independent of the angle θ; we have therefore
$$\frac{dx''}{z'} = -kd\theta:$$
z' differing only by an indefinitely small quantity from z; moreover as dx'' forms with dx the angle $\frac{\pi}{2} - \theta$, we have
$$dx'' = dx \cdot \varphi\left(\frac{\pi}{2} - \theta\right) = \frac{ydx}{z};$$
therefore
$$d\theta = \frac{-ydx}{k \cdot z^2}.$$

If we increase the force y by dy, supposing x constant; we shall have the corresponding variation of the angle θ by changing x into y, y into x, and θ into $\frac{\pi}{2} - \theta$, in the preceding equation, which then gives $d\theta = \frac{xdy}{k \cdot z^2}$; by making x and y vary at the same time, the whole variation of the angle θ will be $\frac{xdy - ydx}{k \cdot z^2}$, and we shall have
$$\frac{xdy - ydx}{z^2} = kd\theta.$$

By substituting for z^2 its value $x^2 + y^2$ and* integrating

---

* The integral or fluent of the quantity $\frac{xdy - ydx}{x^2 + y^2} = kd\theta$, may easily be found by substituting ux for y, and $udx + xdu$ for dy, which gives
$$\frac{xudx + x^2 du - uxdx}{x^2 + u^2 x^2} = \frac{du}{1 + u^2} = kd\theta;$$
therefore an arc whose tang. is u, is equal to $k\theta + \rho$, consequently $u = \frac{y}{x} = $ tang. $(k\theta + \rho)$. In the above it is hardly necessary to observe, that dx, dy, du, and $d\theta$ represent $\dot{x}$, $\dot{y}$, $\dot{u}$, and $\dot{\theta}$, the fluxions of x, y, u, and θ.

we shall have $\frac{y}{x} =$ tang. $(k\theta + \rho)$, ρ being a constant quantity. This equation being combined with that of $x^2 + y^2 = z^2$, gives $x = z \cdot \cos. (k\theta + \rho)$.

It is now only required to find the two constant quantities k and ρ; but if we suppose y to vanish, then evidently $z = x$, and $\theta = 0$; therefore cos. $\rho = 1$, and $x = z$. cos. $k\theta$. If we suppose x to vanish, then $z = y$ and $\theta = \frac{1}{2}\pi$; cos. $k\theta$ being then equal to nothing, k ought to be equal to $2n+1$, n being a whole number, and in this case x will vanish whenever θ shall be equal to $\frac{\frac{1}{2}\pi}{2n+1}$; but x being nothing, we have evidently $\theta = \frac{1}{2}\pi$; therefore $2n+1 = 1$ or $n = 0$, consequently

$$x = z \cdot \cos. \theta.$$

From which it follows, that the diagonal of the rectangle constructed upon the right lines which represent the two forces x and y, represents, not only the quantity, but likewise the direction of their resultant. In like manner we are able for any force whatever to substitute two other forces, which form the sides of a rectangle having that force for the diagonal; and from thence it is easy to conclude, that it is possible to resolve a force into three others which form the sides of a rectangular parallelepiped of which it is the diagonal *.

---

* For if MA (*fig. 2.*) represent any force, it may be resolved into two others, MB and MF, by means of the rectangular parallelogram $MBAF$, also MF may in like manner be resolved into the forces MG and ME, by forming the rectangular parallelogram $MGFE$; then if a parallelepiped be constructed having $MEFG$ and MB for its base and altitude, the force represented by its diagonal will have been resolved into three other forces represented in quantity and direction by its three edges MG, ME, and MB. These three lines are called the co-ordinates of the line MA.

Let therefore a, b, c be the three rectangular co-ordinates of the extremity of the right line, which represents any force whatever, of which the origin is that of the co-ordinates; this force will be represented by the function $\sqrt{a^2+b^2+c^2}$, and by resolving it parallel to the axes of a, b, and c, the partial forces will be respectively represented by these co-ordinates.

Let a', b', c' be the co-ordinates of a second force; $a+a'$, $b+b'$, $c+c'$ will be the co-ordinates of the resultant of the two forces, and will represent the partial ones into which it can be resolved parallel to three axes; from which it is easy to conclude, that this resultant is the diagonal of the parallelogram constructed upon the two forces.

In general a, b, c; a', b', c'; &c. being the co-ordinates of any number whatever of forces; $a+a'+a''$ &c.; $b+b'+b''$ &c; $c+c'+c''$ &c. will be the co-ordinates of their resultant; the square of which will be the sum of the squares of these last co-ordinates; we shall therefore, by this means, have both the magnitude and the position of the resultant.

2. From any point whatever of the direction of a force S, which point we shall take for the origin of this force, let us draw to the material point M a right line which we will call s; let $x, y,$ and z be the three rectangular co-ordinates, which determine the position of the point M, and $a, b,$ and c the co-ordinates of the origin of the force; we shall then have *

$$s = \sqrt{(x-a)^2+(y-b)^2+(z-c)^2}.$$

---

* In (*fig.* 3.) let Ax, Ay, and Az represent the three rectangular co-ordinates of x, y, and z, and MS the line s; from the points S and M let fall the perpendiculars Sm and MN upon the plane yAx, join m and N, draw SR perpendi-

If we resolve the force S parallel to the axes of x, of y, and of z; the corresponding partial forces will be by the preceding n°. $S.\frac{x-a}{s}$, $S.\frac{y-b}{s}$, and $S.\frac{z-c}{s}$, *

---

cular to MN, from m and N, in the plane yAx, draw the perpendiculars mP and NT to Ax, from m draw mQ perpendicular to NT; then because Sm and MN are perpendicular to the same plane, they are parallel to each other; also as mN meets MN in the plane yAx, it is perpendicular to it, and parallel and equal to SR, as is Sm to RN; in the rectangular figure $PTQm$ we have $PT=mQ$, and $Pm=TQ$. In the figure $SM=s$, $Sm=RN=c$, $MN=z$, $MR=MN-NR=z-c$, $NT=y$, $mP=TQ=b$, $NQ=NT-TQ=y-b$, $AT=x$, $AP=a$, $PT=mQ=AT-AP=x-a$, and as MRS is a right angled triangle $MS=\sqrt{SR^2+MR^2}$ but $SR^2=mN^2=$, as the triangle mQN is right angled, $mQ^2+QN^2=PT^2+QN^2$, therefore

$$MS=\sqrt{PT^2+QN^2+MR^2}$$

or by substitution, $s=\sqrt{(x-a)^2+(y-b)^2+(z-c)^2}$. If S coincides with A, then a, b, and c vanish, and

$$s=\sqrt{x^2+y^2+z^2}.$$

Let α, β, and γ respectively represent the angles which s in this case makes with the axes of x, y, and z, then it is evident from fig. 2. in which we may suppose $MA=s$, $MG=x$, $ME=y$, and $MB=z$, that we have the following proportion $s : x : :$ rad. (1) : cos. α, consequently cos. $\alpha=\frac{x}{s}$; in like manner cos. $\beta=\frac{y}{s}$, and cos. $\gamma=\frac{z}{s}$; if these cosines be substituted for their values in the equation

$$s=\sqrt{x^2+y^2+z^2},$$

it will be changed into the following,

$$1=\sqrt{\cos.^2\alpha+\cos.^2\beta+\cos.^2\gamma}.$$

* By the preceding number $s : x-a : : S : S.\frac{x-a}{s}$ as the force in the direction of the axis x.

or $S.\left(\frac{\delta s}{\delta x}\right)$; $S.\left(\frac{\delta s}{\delta y}\right)$; and $S.\left(\frac{\delta s}{\delta z}\right)$; $\left(\frac{\delta s}{\delta x}\right)$, $\left(\frac{\delta s}{\delta y}\right)$, and $\left(\frac{\delta s}{\delta z}\right)$* denoting, according to the received notation, the co-efficients of the variations δx, δy, and δz, in the variation of the preceding expression of s.

If, in like manner, we name s' the distance of M from any point in the direction of another force S', that point being taken for the origin of the force; $S'.\left(\frac{\delta s'}{\delta x}\right)$ will be this force resolved parallel to the axis of x, and so on of the rest; the sum of the forces S, S', S'', &c. resolved parallel to this axis will therefore be $\Sigma. S.\left(\frac{\delta s}{\delta x}\right)$ the characteristic Σ of finite integrals denoting here, the sum of the terms $S.\left(\frac{\delta s}{\delta x}\right)$, $S'.\left(\frac{\delta s'}{\delta x}\right)$, &c.

Let V be the resultant of all the forces S, S', &c., and u the distance of the point M from a point in the

---

* The expressions $\left(\frac{\delta s}{\delta x}\right)$, $\left(\frac{\delta s}{\delta y}\right)$, $\left(\frac{\delta s}{\delta z}\right)$ enclosed between parentheses, represent the co-efficients of the partial differentiations of the equation
$$s = \sqrt{(x-a)^2 + (y-b)^2 + (z-c)^2}$$
taken by making x, y, and z vary separately; thus differentiating the equation, by supposing y and z constant, we obtain $\delta s = \frac{(x-a)\delta x}{s}$ therefore $\left(\frac{\delta s}{\delta x}\right) = \frac{x-a}{s}$, in like manner, $\left(\frac{\delta s}{\delta y}\right) = \frac{y-b}{s}$ and $\left(\frac{\delta s}{\delta z}\right) = \frac{z-c}{s}$: these expressions are evidently equivalent to the co-sines of the angles which the line s makes with the co-ordinates x, y, and z respectively.

direction of this resultant, which is taken for its origin; $V.\left(\frac{\delta u}{\delta x}\right)$ will be the expression of this resultant resolved parallel to the axis of x; we shall therefore have by the preceding number $V.\left(\frac{\delta u}{\delta x}\right) = \Sigma . \, S.\left(\frac{\delta s}{\delta x}\right)$.

We shall have in like manner,

$$V.\left(\frac{\delta u}{\delta y}\right) = \Sigma . \, S.\left(\frac{\delta s}{\delta y}\right). \qquad V.\left(\frac{\delta u}{\delta z}\right) = \Sigma . \, S.\left(\frac{\delta s}{\delta z}\right).$$

from which we may obtain, by multiplying these three equations respectively by δx, δy, and δz, and then adding them together,

$$V. \, \delta u = \Sigma . \, S . \, \delta s ; \quad \ldots \quad (a)$$

As this last equation has place, whatever may be the variations δx, δy, and δz, it is equivalent to the three preceding ones.

If its second member is an exact variation of a function φ, we shall have $V. \, \delta u = \delta \varphi$, and consequently,

$$V.\left(\frac{\delta u}{\delta x}\right) = \left(\frac{\delta \varphi}{\delta x}\right);$$

that is to say, the sum of all the forces S, S', &c. resolved parallel to the axis of x is equal to the partial differential $\left(\frac{\delta \varphi}{\delta x}\right)$. This case generally takes place when these forces are respectively functions of the distance of their origin from the force M. In order to have the resultant of all these forces resolved parallel to any right line whatever, we shall take the integral $\Sigma . \int . S \, \delta s$, and naming it φ, we will consider it as a function of x, and of two other right lines perpendicular to x and to each other; the partial differential $\left(\frac{\delta \varphi}{\delta x}\right)$ will then be the resultant of the forces S, S', &c. resolved parallel to the right line x.*

---

* The following expressions of the equilibrium of a point follow from the above equations. Suppose that the powers are represented both in magnitude and direction by S, S', S'', &c. whose directions form the following angles,

with the axis of x . . . $\alpha, \alpha', \alpha'',$ . . .
with the axis of y . . . $\beta, \beta', \beta'',$ . . .
with the axis of z . . . $\gamma, \gamma', \gamma'',$ . . .

By resolving each of these forces into three others whose directions are parallel to the axes, we shall have for the composing forces parallel

to x . . $S.\cos.\alpha$, $S'.\cos.\alpha'$, $S''.\cos.\alpha''$, &c. .
to y . . $S.\cos.\beta$, $S'.\cos.\beta'$, $S''.\cos.\beta''$, &c. .
to z . . $S.\cos.\gamma$, $S'.\cos.\gamma'$, $S''.\cos.\gamma''$, &c. .

Each of these three collections of forces is equivalent to a single force, equal to their sum, because these components are directed in the same right line. Naming P, Q, and R the three forces respectively parallel to x, y, and z, we shall have

$P = S.\cos.\alpha + S'.\cos.\alpha' + S''.\cos.\alpha'' + $&c.
$Q = S.\cos.\beta + S'.\cos.\beta' + S''.\cos.\beta'' + $&c.
$R = S.\cos.\gamma + S'.\cos.\gamma' + S''.\cos.\gamma'' + $&c.

Let a, b, and c represent the unknown angles which the direction of the resultant V forms with the three axes; $V.\cos.a$, $V.\cos.b$, $V.\cos c$ will be its components in the directions of the axes; we shall have therefore $V.\cos.a = P$, $V.\cos.b = Q$, and $V.\cos.c = R$. If we add the squares of these equations together, remembering that $\cos.^2 a + \cos.^2 b + \cos.^2 c = 1$, we shall obtain $V^2 = P^2 + Q^2 + R^2$, which gives $V = \sqrt{(P^2 + Q^2 + R^2)}$; the direction of the resultant may be obtained from the equations

$$\cos. a = \frac{P}{V}, \quad \cos. b = \frac{Q}{V}, \quad \cos. c = \frac{R}{V}.$$

These equations determine both the magnitude and the di-

3. When the point M is in equilibrio by the action of all the forces which solicit it, their resultant is nothing, and the equation *(a)* becomes

$$0 = \Sigma . S . \delta s ; \quad (b)$$

which shews, that in the case of the equilibrium of a point acted upon by any number whatever of forces, the sum of the products of each force by the element of its direction is nothing.

If the point M is forced to be upon a curved surface, it will experience a re-action which we shall denote by R. This re-action is equal and directly contrary to the pressure with which the point presses upon the surface; for by supposing it acted upon by two forces, R and $-R$, it is possible to conceive, that the force $-R$ is destroyed by the re-action of the surface, and that the point M presses upon the surface with the force $-R$; but the force of pressure of a point upon a surface is perpendicular to it, otherwise it would be possible to resolve the force into two, one perpendicular to the surface, which would be destroyed by it, the other parallel to the surface, in consequence of which the point would have no action upon it, which is contrary to the

---

rection of the resultant V, which is evidently the diagonal of the parallelepiped constructed upon P, Q, and R. If the system be in equilibrio, it is manifest that each collection of forces parallel to the axes should likewise be in equilibrio, which gives

$$P = 0, \quad Q = 0, \quad R = 0.$$

With respect to the signs of the components $S . \cos . \alpha$, S', $\cos . \alpha'$, &c., it may be observed that those which tend to increase the co-ordinates should be reckoned positive, and those which act in a contrary direction negative.

supposition; naming *r therefore the perpendicular drawn from the point M at the surface, and terminated in any point whatever of its direction, the force R will be directed along this perpendicular; it will be necessary therefore to add $R.\delta r$ to the second member of the equation *(b)* which will become

$$0 = \Sigma . S . \delta s + R . \delta r ; \quad (c)$$

—R being then the resultant of all the forces S, S', &c it is perpendicular to the surface.

If we suppose that the arbitrary variations $\delta x, \delta y$, and δz appertain to the curve surface upon which the point is forced to remain, we shall have, by the nature of the perpendicular to this surface, $\delta r = 0$, which makes $R.\delta r$ vanish from the preceding equation: the equation *(b)* has place therefore in this case, provided that we extract one of the three variations $\delta x, \delta y$, and δz, by means of the equation to the surface; but then the equation *(b)* which in the general case is equivalent to three, is not equivalent to more than two distinct equations, which we may obtain by equalling separately to nothing the co-efficients of the two remaining differentials. † Let $u = 0$ be the equation of the surface,

---

* In *(fig. 4.)* let a point be in equilibrio at M, on the curve AMB, by means of the forces MP, MQ, and the reaction of the curve; then if MR, supposed perpendicular to the curve at M, be the resultant of the forces MP, and MQ, it will represent the pressure of the point upon the curve, and if RM be produced to r, and $Mr = MR$, then Mr will represent the re-action of the curve upon the point, which may be supposed in equilibrio in consequence of the forces MP, MQ, and Mr.

† The nature of a surface may be determined by three rectangular co-ordinates, as that of a line may by two; thus let

the two equations $\delta r=0$ and $\delta u=0$ will have place at the same time, this requires that δr should be equal to $N\delta u$, N being a function of x, y, and z. Naming a,

---

$u=0$ be an equation to a surface, and let its co-ordinates x, y, and z be respectively measured upon the lines AX, AY, and AZ (*fig.* 5.); if the values of x and y are given and represented by a and b, by taking on the axes of x and y $AP=a$, $AQ=b$, and drawing the parallels QM', PM' to these axes, the point M', which is the projection of the point M of the surface upon the plane of x, y, will be determined; the equation by substitution will then give the corresponding value of z, which determines the length of the co-ordinate $M'M$, and, consequently the point M of the surface.

Curves of double curvature are formed by the intersection of two surfaces. Thus, let the equations to two surfaces be represented by $F(x, y, z)=0$, and $f(x, y, z)=0$, then the curve of double curvature formed by their intersection will have the same co-ordinates; if therefore the variable x be extracted by means of these equations, the resulting one will represent the projection of the curve of double curvature upon the plane of yz; in like manner, if y had been extracted, the resulting one would have been that of the projection upon the plane of xz; and if z had, that of the projection upon the plane of xy. The resulting equations likewise represent cylindrical surfaces elevated upon these projections respectively perpendicular to the planes of the co-ordinates, and the curve of double curvature will be the intersection of any two of these surfaces.

Those who are desirous of being acquainted with the properties of surfaces, or of curves of double curvature, or of lines supposed in space, and considered with reference to their projections upon three planes at right angles to each other, may consult the Traite' du Calcul Differentiel et Integral par S. F. Lacroix, the Application de l'Algebre a la Geometrie par MM. Monge et Hachette, &c.

b, and c the co-ordinates of the origin of r, we shall have to determine it

$$r = \sqrt{(x-a)^2 + (y-b)^2 + (z-c)^2};$$

from which we may obtain

$$\left(\frac{\delta r}{\delta x}\right)^2 + \left(\frac{\delta r}{\delta y}\right)^2 + \left(\frac{\delta r}{\delta z}\right)^2 = 1; \text{ and consequently}$$

$$N \cdot \left\{\left(\frac{\delta u}{\delta x}\right)^2 + \left(\frac{\delta u}{\delta y}\right)^2 + \left(\frac{\delta u}{\delta z}\right)^2\right\} = 1;$$

by making therefore

$$\lambda = \frac{R}{\sqrt{\left(\frac{\delta u}{\delta x}\right)^2 + \left(\frac{\delta u}{\delta y}\right)^2 + \left(\frac{\delta u}{\delta z}\right)^2}};$$

the term $R \cdot \delta r$ of the equation *(c)* will be changed into $\lambda \delta u$, and this equation will become

$$0 = \Sigma \cdot S \cdot \delta s + \lambda \cdot \delta u;$$

in which we ought to equal separately to nothing the co-efficients of the variations δx, δy, and δz, which gives three equations; but they are only equivalent to two between x, y, and z, on account of the indeterminate quantity λ which they contain. We may therefore instead of extracting from the equation *(b)* one of the variations δx, δy, or δz, by means of the differential equation to the surface, add to it this equation multiplied by an indeterminate quantity λ, and then consider the variations δx, δy, and δz as independant quantities. This method, which likewise results from the theory of elimination, unites to the advantage of simplifying the calculation, that of making known the pressure $-R$ with which the point M acts against the surface*.

---

* Let the point be supposed in equilibrio upon a surface whose equation is $u = 0$, and let the forces S, S', S'', &c. be reduced to three P, Q, and R acting in the directions of the

. Supposing this point to be contained in a canal of simple or double curvature, it will prove on the part

---

three rectangular co-ordinates; then the sum of the moments $P.\delta x + Q.\delta y + R.\delta z$ will be equivalent to $S\delta s + S'\delta s' + S''\delta s''$ +&c. and by adding δu multiplied by the indeterminate quantity λ, the equation of equilibrium becomes
$$P.\delta x + Q.\delta y + R.\delta z + \lambda \delta u = 0;$$
but u being a known function of x, y, and z, we shall have by differentiation
$$\delta u = \left(\frac{\delta u}{\delta x}\right)\delta x + \left(\frac{\delta u}{\delta y}\right)\delta y + \left(\frac{\delta u}{\delta z}\right)\delta z,$$
$\left(\frac{\delta u}{\delta x}\right)$, $\left(\frac{\delta u}{\delta y}\right)$, and $\left(\frac{\delta u}{\delta z}\right)$ representing the co-efficients of δx, δy, and δz. By substituting this value of δu in the preceding equation, it becomes
$$P.\delta x + Q.\delta y + R.\delta z + \lambda\left(\frac{\delta u}{\delta x}\right)\delta x + \lambda\left(\frac{\delta u}{\delta y}\right)\delta y + \lambda\left(\frac{\delta u}{\delta z}\right)\delta z = 0$$
which gives, by equalling separately to nothing each sum of the terms multiplied respectively by δx, δy, and δz, the three following equations
$$P + \lambda.\left(\frac{\delta u}{\delta x}\right) = 0,$$
$$Q + \lambda.\left(\frac{\delta u}{\delta y}\right) = 0,$$
$$R + \lambda.\left(\frac{\delta u}{\delta z}\right) = 0;$$
from which, by extracting λ, we shall obtain the two following equations
$$Q.\left(\frac{\delta u}{\delta x}\right) - P.\left(\frac{\delta u}{\delta y}\right) = 0,$$
$$R.\left(\frac{\delta u}{\delta x}\right) - P.\left(\frac{\delta u}{\delta z}\right) = 0;$$
that contain the conditions of the equilibrium of a point upon a surface.

of the canal, a re-action which we shall denote by k, that will be equal and directly contrary to the pressure with which the point acts against the canal, the direction of which will be perpendicular to its side; but the curve formed by this canal is the intersection of two

---

In the case of a point acted upon by certain forces, the conditions of its equilibrium upon a surface may be found with more ease, by directly substituting in the equation
$$P.\delta x + Q.\delta y + R.\delta z = 0,$$
the value of δz obtained from the differential equation
$$\left(\frac{\delta u}{\delta x}\right)\delta x + \left(\frac{\delta u}{\delta y}\right)\delta y + \left(\frac{\delta u}{\delta z}\right)\delta z = 0,$$
of the surface, and then equalling separately to nothing the co-efficients of the differentials δx and δy. By this method we shall immediately get the equations
$$P - R.\frac{\left(\frac{\delta u}{\delta x}\right)}{\left(\frac{\delta u}{\delta z}\right)} = 0,$$
$$Q - R.\frac{\left(\frac{\delta u}{\delta y}\right)}{\left(\frac{\delta u}{\delta z}\right)} = 0;$$
which are equivalent to the equations found by the other method.

In like manner, if a body is forced to be upon a line of a given description, determined by the two differential equations $\delta y = p\delta x$, $\delta z = q\delta x$ of the projections of the line upon the planes of xy and xz, we have only to substitute these values of δy and δz, in the equation $P.\delta x + Q.\delta y + R.\delta z = 0$, which, on being divided by δx, gives the equation,
$$P. + Q.p + R.q = 0,$$
for the condition of equilibrium.

surfaces, of which the equations express its nature; we may therefore consider the force k as the resultant of the two forces R and R', which re-act upon the point M from the two surfaces; moreover as the directions of the three forces R, R', and k are perpendicular to the side of the curve, they are in the same plane. By naming therefore δr and $\delta r'$ the elements of the directions of the forces R and R', which directions are respectively perpendicular to each surface, it will be necessary to add to the equation *(b)* the two terms $R.\delta r$ and $R'.\delta r'$ which will change it into the following

$$0 = \Sigma . S . \delta s + R . \delta r + R' . \delta r'. \quad (d)$$

If we determine the variations δx, δy, and δz so that they may appertain at the same time to two surfaces, and, consequently, to the curve formed by the canal; δr and $\delta r'$ will vanish, and the preceding equation will be reduced to the equation *(b)*, which therefore has place again in the case where the point M is forced to move in a canal; provided, that by means of the two equations which express the nature of the canal, we make two of the variations δx, δy, and δz to disappear.

Let us suppose that $u = 0$ and $u' = 0$ are the equations of the two surfaces, whose intersection forms the canal. If we make

$$\lambda = \frac{R}{\sqrt{\left(\frac{\delta u}{\delta x}\right)^2 + \left(\frac{\delta u}{\delta y}\right)^2 + \left(\frac{\delta u}{\delta z}\right)^2}};$$

and

$$\lambda' = \frac{R'}{\sqrt{\left(\frac{\delta u'}{\delta x}\right)^2 + \left(\frac{\delta u'}{\delta y}\right)^2 + \left(\frac{\delta u'}{\delta z}\right)^2}};$$

the equation *(d)* will become

$$0 = \Sigma . S . \delta s + \lambda \delta u + \lambda' \delta u';$$

an equation in which the co-efficients of each of the variations δx, δy, and δz will be separately equal to nothing; in this manner three equations will be obtained, by means of which the values of λ and λ' may be determined, which will give the re-actions R and R' of the two surfaces; and by composing them we shall have the re-action k of the canal upon the point M, and, consequently, the pressure with which this point acts against the canal. This re-action, resolved parallel to the axis of x is equal to

$$R.\left(\frac{\delta r}{\delta x}\right)+R'.\left(\frac{\delta r'}{\delta x}\right); \text{ or to } \lambda.\left(\frac{\delta u}{\delta x}\right)+\lambda'.\left(\frac{\delta u}{\delta x}\right);$$

the equations of condition $u=0$, and $u'=0$, to which the motion of the point M is subjected, express, therefore, by means of the partial differentials of functions, which are equal to nothing because of these equations, the resistances which act upon this point in consequence of the conditions of its motion.

It appears from what precedes, that the equation *(b)* of equilibrium has generally place, provided, that the variations δx, δy, and δz are subjected to the conditions of equilibrium. This equation may therefore be made the foundation of the following principle.

If an indefinitely small variation is made in the position of the point M, so that it still remains upon the surface, or the curve along which it would move if it were not entirely free; the sum of the forces which solicit it, each multiplied by the space which the point moved in its direction, is equal to nothing in the case of equilibrium *.

---

* Let the forces S, S', S'', &c. be supposed to act upon the point M in the directions of the lines s, s', s'', &c. re-

The variations δx, δy, and δz being supposed arbitrary and independent, it is possible in the equation

spectively drawn from that point to the origins' of these forces; let V represent their resultant, and u a line drawn from the point M in its direction; also, let any line MN, be supposed to be drawn from the point M, and each of the forces V, S, S', &c. to be resolved into two others, one in the direction of this line, and the other perpendicular to it. Because V is the resultant of all the other forces, its component along the line MN will be equal to the sum of the components of the other forces along the same line; let m, a, a', a'', &c. denote the angles which the directions of the forces V, S, S', S'', &c. respectively make with the line MN, we shall then have the following equation,

$$V.\cos.m = S.\cos.a + S'.\cos.a' + S''.\cos.a'' + \&c.$$

Let any point N be taken upon the line MN, then if this line be represented by b, and its respective projections upon the lines m, s, s', s'', &c. or their continuations by Δu, Δs, $\Delta s'$, $\Delta s''$, &c. we shall have

$$\Delta u = b.\cos.m, \quad \Delta s = b.\cos.a, \quad \Delta s' = b.\cos.a', \&c.$$

If both sides of the preceding equation are multiplied by b, it will, by substitution, become

$$V.\Delta u = S.\Delta s + S'.\Delta s' + S''.\Delta s'' + \&c.$$

If the point M be supposed to move to the point N, and the line MN be regarded as representing the virtual velocity of this point, the quantities Δu, Δs, $\Delta s'$, &c. will denote the virtual velocities of this point in the directions of the forces V, S, S', S'', &c.; the last equation therefore shews, that the product of the resultant of any number of forces applied to the same point, by the virtual velocity of this point estimated in its direction, is equal to the sum of the products of these forces by their respective virtual velocities, estimated in the directions of the forces. It is not absolutely necessary that

(a) to substitute for the co-ordinates x, y and z three other quantities which are functions of them, and to equal the variations of these quantities to nothing.

Thus naming ρ the radius drawn from the origin of the co-ordinates to the projection of the point M upon the plane of x and y, and ϖ the angle formed by ρ and the axis of x, we shall have

$$x = \rho.\cos.\varpi\ ;\quad y = \rho.\sin.\varpi.$$

By considering therefore in the equation *(a)*, u, s, s', &c. as functions of ρ, ϖ, and z, and comparing the coefficients of $\delta\varpi$, we shall have

$$V.\left(\frac{\delta u}{\delta \varpi}\right) = \Sigma.S.\left(\frac{\delta s}{\delta \varpi}\right)\ ;\quad (e)$$

$\frac{V}{\rho}.\left(\frac{\delta u}{\delta \varpi}\right)$ is the expression of the force V resolved in the direction of the element $\rho.\delta\varpi$. Let V' be this force resolved parallel to the plane of x and y, and p the perpendicular let fall from the axis of z, upon the direction

---

the virtual velocities should be supposed indefinitely small, if they are, the last equation becomes

$$V.\delta u = S.\delta s + S'.\delta s' + S''.\delta s'' + \&c.$$

In the case of equilibrium, we have $V=0$, and consequently

$$0 = S.\delta s + S'.\delta s' + S''\delta s'',\ \&c.$$

If the point is forced to remain upon a given curve or surface, this equation will be proper, the virtual velocities δs, $\delta s'$, $\delta s''$, &c. being supposed indefinitely small, and $\delta u = 0$ instead of V.

The products $S.\delta s$, $S'.\delta s'$, $S''.\delta s''$, &c. are called by some authors the moments of the powers S, S', S'', &c.; and their sum being equal to nothing, which has place not only for one point, but, as will be proved hereafter, for any system whatever in equilibrio, is called the principle of virtual velocities.

of V' parallel to the same plane; $\dfrac{pV'}{\rho}$ will be a second expression of the force V resolved in the direction of the element $\rho\delta\varpi$ *; we shall therefore have

$$pV' = V \cdot \left(\dfrac{\delta u}{\delta \varpi}\right).$$

If we suppose the force V' to be applied to the extremity of the perpendicular p, it will tend to make it turn about the axis of z; the product of this force by the perpendicular is what is called the moment of the force V with respect to the axis of z; this moment is therefore equal to $V \cdot \left(\dfrac{\delta u}{\delta \varpi}\right)$; and it appears from the equation (e), that the moment of the resultant of any number whatever of forces is equal to the sum of the moments of these forces

---

* Let MB (*fig.* 6.) represent V', or the force V resolved along a plane parallel to that of xy, let O be the point where this plane cuts the axis of z, join OM, then OM will be parallel and equal to ρ, draw OA or p perpendicular to V' or MB produced, also BD perpendicular to OM; then, as the right angled triangles MBD, MOA have a common angle at M, they are similar, consequently

$$MO(\rho) : OA(p) :: MB(V') : DB = \dfrac{pV'}{\rho};$$

but the line DB represents the force V' resolved in the direction $\rho\delta\varpi$ perpendicular to ρ, therefore that force is represented by $\dfrac{pV'}{\rho}$.

CHAP. II.

Of the motion of a material point.

4. A Point at rest is not able to give itself any motion, because it does not contain within itself any cause why it should move in one direction rather than another. When it is solicited by any force whatever, and afterwards left to itself, it will move constantly in an uniform manner in the direction of that force, if not opposed by any resistance. This tendency of matter to persevere in its state of motion or rest, is called its inertia. This is the first law of the motion of bodies.

That the direction of the motion is a right line follows evidently from this, that there is not any reason why the point should change its course more to one side than the other of its first direction: but the cause of the uniformity of its motion is not so evident. The nature of the moving force being unknown, it is impossible to discover *a priori* if it ought to continue without ceasing. In fact, as a body is incapable of giving itself any motion, it appears equally incapable of altering that which it has received; so that the law of inertia is the most simple and natural which it is possible to imagine; it is also confirmed by experience, for we observe upon the earth that the more the obstacles are diminished, which oppose the motions of bodies, the longer these motions are continued; which leads us to believe, that if the obstacles were removed they would never cease. But the inertia of matter is most remarkable in the motions of the celestial bodies, which have not during a great number of ages experienced any perceptible alteration. Thus we may regard the inertia of bodies as a law of nature, and when we shall observe any alteration in the motion of a body, we will suppose that it is owing to the action of a different cause.

In uniform motion the spaces gone over are in proportion to the times, but the time employed in describing a given space, is longer or shorter according to the magnitude of the moving force. These differences have given rise to the notion of velocity, which, in uniform motion, is the ratio of the space to the time passed in going over it: thus, s representing the space, t the time, and v the velocity, we have $v = \frac{s}{t}$. Time and space being heterogeneal, and consequently, not comparable quantities, a determinate interval of time is chosen, such as a second for an unit of time; in like manner, some unit of space is chosen, as a metre; and then space and time become abstract numbers, which express how often they contain the units of their species, that are thus rendered comparable to each other. By this means the velocity becomes the ratio of two abstract numbers, and its unity is the velocity of a body which passes over the space of a metre in one second.

5. The force being only known by the space which

it causes a body to describe in a given time, it is natural to take this space for its measure ; but this supposes that many forces acting in the same direction, should cause the body to pass over a space equal to the sum of the spaces which each of them would have made it go over separately ; or what comes to the same, that the force is proportional to the velocity. This is what we are not able to know *a priori*, owing to our ignorance of the nature of the moving force ; it is therefore necessary to have again recourse to experience on this occasion, for all that which is not a necessary consequence of the little which we know respecting the nature of things, must be to us but a result of observation*.

---

* Mr. Knight, in the ninth No. of the Mathematical Repository, has attempted to prove the law of the proportionality of the force to the velocity, by supposing two straight lines at right angles to each other, to represent the magnitudes and the directions of two forces, and taking parts from these lines, measured from their junction, to represent the velocities which these forces would respectively cause ; and by completing two parallelograms, one about the lines denoting the forces, and the other about those denoting the velocities, and drawing diagonals in each of them representing the respective resultants of the forces and the velocities ; he has shewn that if the diagonals are in the same right line, the parallelograms will be similar, and consequently the forces and the respective velocities proportional ; if they are not, it must be supposed that the resultant of the forces has caused a motion in a different direction to its own, which is absurd.

Mr. Knight has, I think, proved that the force varies as the velocity, if it be taken for granted, that the proofs re-

Naming v the velocity of the earth, which is common to all the bodies upon its surface, let f be the force by which one of these bodies M is actuated in consequence of this velocity, and let us suppose that $v = f\varphi$ (f) is the relation which exists between the velocity and the force; φ (f) being a function of f which it is necessary to determine by experiment. Let a, b, and c be the three partial forces into which the force f may be resolved parallel to three axes which are perpendicular to each other. Let us then suppose the moving body M to be solicited by a new force f', which may be resolved into three others a', b', and c' parallel to the same axes. The forces by which this body will be actuated in the directions of these axes, are $a+a'$, $b+b'$, and $c+c'$; naming F the sole resulting force, it will become, from what precedes,

$$F = \sqrt{(a+a')^2 + (b+b')^2 + (c+c')^2}.$$

If the velocity corresponding to F be named U; $\dfrac{(a+a').U}{F}$* will represent this velocity resolved pa-

---

specting the composition and the resolution of forces, and those respecting the composition and the resolution of velocities, are satisfactorily demonstrated independent of each other.

* If U represents the velocity of the body corresponding to F, we shall find that part of it relative to the axis of $a+a'$ by the proportion

$$\sqrt{(a+a')^2+(b+b')^2+(c+c')^2} : a+a' :: U : \dfrac{(a+a')U}{\sqrt{(a+a')^2+(b+b')^2+(c+c')^2}};$$

but $F = \sqrt{(a+a')^2+(b+b')^2+(c+c')^2}$, therefore, by substitution the expression becomes $\dfrac{(a+a')U}{F}$.

rallel to the axis of a; also the relative velocity of the body upon the earth parallel to this axis will be $\frac{(a+a')U}{F} - \frac{av}{f}$, or $(a+a') \cdot \varphi(F) - a \cdot \varphi(f)$. The greatest forces which we are able to impress upon bodies at the surface of the earth, being much smaller than those by which they are actuated in consequence of the motion of the earth, we may consider a', b', and c' as indefinitely small quantities relative to f: we shall therefore have $F = f + \frac{aa' + bb' + cc'}{f}$ *; and

---

* If a', b', and c' are supposed indefinitely small relative to f, their squares and products may be neglected: the expression $\sqrt{(a+a')^2 + (b+b')^2 + (c+c')^2}$ will then become $\sqrt{a^2 + b^2 + c^2 + 2aa' + 2bb' + 2cc'}$; let $a^2 + b^2 + c^2 = f^2$ and $2aa' + 2bb' + 2cc' = y$, then if the expression $\sqrt{f^2 + y}$ is expanded by the binomial theorem, it will become $f + \frac{y}{2f}$ &c. or $f + \frac{aa' + bb' + cc'}{f}$ &c. but as all the terms after the two first of the series contain the squares, products, or higher powers of a', b', and c, they may be neglected.

If in the expression $\varphi(x)$ we substitute $x + k$ for x, it becomes $\varphi(x+k) = \varphi(x) + \frac{d\varphi(x)}{dx} \cdot k + \frac{d^2\varphi(x)}{1.2 dx^2} \cdot k^2 + \frac{d^3\varphi(x)}{1.2.3 dx^3} \cdot k^3 + \frac{d^4\varphi(x)}{1.2.3.4 dx^4} \cdot k^4 +$ &c. which gives, if $\varphi(x)$ be represented by u, $\varphi(x+k) = u + \frac{du}{dx} \cdot k + \frac{d^2u}{1.2. dx^2} \cdot k^2 + \frac{d^3u}{1.2.3 dx^3} \cdot k^3 +$ &c this is called the theorem of Taylor, and is proved a variety of ways in different mathematical works. If f be substituted

$$\varphi(F) = \varphi(f) + \frac{aa' + bb' + cc'}{f} \cdot \varphi'(f) \, ; \quad \varphi'(f) \text{ being the}$$

differential of $\varphi(f)$ divided by df. The relative velocity of M in the direction of the axis of a, will, in like manner, become

$$a' \cdot \varphi(f) + \frac{a}{f} \cdot \{aa' + bb' + cc'\} \cdot \varphi(f).$$

Its relative velocities in the directions of the axes b and c will be

$$b' \cdot \varphi(f) + \frac{b}{f} \cdot \{aa' + bb' + cc'\} \cdot \varphi'(f) \, ;$$

$$c' \cdot \varphi(f) + \frac{c}{f} \cdot \{aa' + bb' + cc'\} \cdot \varphi'(f).$$

The position of the axes $a, b,$ and c being arbitrary, we may take the direction of the impressed force for the axis of a, and then b' and c' will vanish; and the preceding relative velocities will be changed into the following,

$$a' \cdot \left\{ \varphi(f) + \frac{a^2}{f} \cdot \varphi'(f) \right\} \, ; \quad \frac{ab}{f} \cdot a' \cdot \varphi'(f) \, ; \quad \frac{ac}{f} \cdot a' \cdot \varphi'(f).$$

If $\varphi'(f)$ does not vanish, the moving body in consequence of the impressed force a', will have a relative velocity perpendicular to the direction of this force, provided that b and c do not vanish; that is to say, provided that the direction of this force does not coin-

---

for x, and $\dfrac{aa' + bb' + cc'}{f}$ for k in the above expression, it becomes

$$\varphi\left(f + \frac{aa' + bb' + cc'}{f}\right) = \varphi(f) + \frac{aa' + bb' + cc'}{f} \cdot \varphi'(f) + \&c.\, ;$$

but as the quantities a', b', and c' are indefinitely small, the products and higher powers than the first of $\dfrac{aa' + bb' + cc'}{f}$ may be neglected.

side with that of the motion of the earth. Thus, suppose that a globe at rest upon a very smooth horizontal plane is struck by the base of a right angled cylinder, moving in the direction of its axis, which is supposed horizontal, the apparent relative motion of the globe will not be parallel to this axis in all its positions with respect to the horizon : thus we have an easy means of discovering by experiment if $\varphi'(f)$ has a perceptible value upon the earth ; but the most exact experiments have not shewn in the apparent motion of the globe any deviation from the direction of the impressed force ; from which it follows that upon the earth, $\varphi'(f)$ is very nearly nothing. If its small value were perceptible, it would particularly be shewn in the duration of the oscillations of the pendulum, which would alter as the position of the plane of its motion differed from the direction of the motion of the earth. As the most exact observations have not discovered any such difference, we ought to conclude that $\varphi'(f)$ is insensible, and may be supposed equal to nothing upon the surface of the earth.

If the equation $\varphi'(f)=0$ has place, whatever the force f may be, $\varphi(f)$ will be constant, and the velocity will be proportional to the force ; it will also be proportional to it if the function $\varphi(f)$ is composed of only one term, as otherwise $\varphi(f)$ would not vanish except f did : it is necessary, therefore, if the velocity is not proportional to the force, to suppose that in nature the function of the velocity which expresses the force is formed of many terms, which is hardly probable ; it is also necessary to suppose that the velocity of the earth is exactly that which belongs to the equation $\varphi'(f)=0$, which is contrary to all probability. Moreover the velocity of the earth varies during the different seasons of the year ; it is about one thirtieth part greater in winter than in summer. This variation is still more considerable, if, as every thing appears to indicate, the solar system be in motion in space; for according as this progressive motion agrees with that of the earth, or is contrary to it, there should result during the course of the year very great variations in the absolute motion of the earth, which would alter the equation that we are considering, and the relation of the impressed force to the absolute velocity which results, if this equation and this ratio were not independent of the motion of the earth; nevertheless, the smallest difference has not been discovered by observation.

Thus we have two laws of motion, the law of inertia, and that of the force being proportional to the velocity, which are given from experience. They are the most natural and the most simple which it is possible to imagine, and are without doubt derived from the nature itself of matter; but this nature being unknown, they are with respect to us solely the consequences of observation, and the only ones which the science of mechanics requires from experience.

6. As the velocity is proportional to the force, these two quantities may be represented by each other, and all that has been previously established respecting the composition of forces, may be applied to the composition of velocities *. It therefore results that the rela-

---

* Let v', v'', and v''' represent the uniform velocities impressed upon a body in the directions of three rectangular co-ordinates, x, y, and z, the spaces respectively passed over in the time t in consequence of them; we shall then have the three following equations

$$x = v't, \quad y = v''t, \quad z = v'''t,$$

tive motions of a system of bodies actuated by any forces whatever, are the same whatever may be their common motion; for this last motion resolved into three others parallel to three fixed axes, causes the partial

---

and the resulting motion will be uniform and rectilinear, and determined by the equation
$$s = \sqrt{(x^2 + y^2 + z^2)} = t\sqrt{(v'^2 + v''^2 + v'''^2)},$$
in which s represents the space gone over. If v represents the velocity of the body, it will be equal to
$$\sqrt{(v'^2 + v''^2 + v'''^2)}.$$
The co-sines of the angles which the direction s of the motion forms with the co-ordinates x, y, and z respectively, are
$$\frac{x}{s} = \frac{v'}{\sqrt{(v'^2 + v''^2 + v'''^2)}}, \quad \frac{y}{s} = \frac{v''}{\sqrt{(v'^2 + v''^2 + v'''^2)}},$$
$$\frac{z}{s} = \frac{v'''}{\sqrt{(v'^2 + v''^2 + v'''^2)}}.$$
Let s, x, y, and z have the same significations as above, and g', g'', and g''' denote the constant accelerating forces impressed parallel to the axes of x, y, and z; then the equations of the motion of the point being
$$x = \tfrac{1}{2}g't^2; \quad y = \tfrac{1}{2}g''t^2; \quad z = \tfrac{1}{2}g'''t^2;$$
the equations of the projections of the line passed over upon the planes of xy and yz, will be, as appears by extracting t^2 from the two first and two last equations
$$y = \frac{g''}{g'}x; \quad z = \frac{g'''}{g'}x;$$
the line passed over will therefore be a right line, and we shall have.
$$s = \sqrt{(x^2 + y^2 + z^2)} = \tfrac{1}{2}t^2\sqrt{(g'^2 + g''^2 + g'''^2)}.$$

velocities of each body parallel to these axes to increase by the same quantity, and as their relative velocity only depends upon the difference of these partial velocities, it is the same whatever may be the motion com-

---

If α, β, and γ represent the co-sines of the angles which s makes with x, y, and z respectively, then

$$\cos.\alpha = \frac{g'}{\sqrt{(g'^2+g''^2+g'''^2)}}, \quad \cos.\beta = \frac{g''}{\sqrt{(g'^2+g''^2+g'''^2)}},$$

$$\cos.\gamma = \frac{g'''}{\sqrt{(g'^2+g''^2+g'''^2)}}.$$

The accelerating force in the direction of s is constant and equal to $\sqrt{(g'^2+g''^2+g'''^2)}$, and composed of the three given accelerating forces, as is the case with uniform motions.

Retaining the foregoing notation, and supposing that v', v'', and v''' represent the initial velocities of a point acted upon by constant accelerating forces parallel to the three axes; the equations of the impressed motion will be

$$x = v't + \tfrac{1}{2}g't^2\,;\ y = v''t + \tfrac{1}{2}g''t^2\,;\ z = v'''t + \tfrac{1}{2}g'''t^2.$$

The projection of the curve passed over by the point upon the plane of xy, found by extracting t from the two first equations, is

$$y^2 + \frac{g''^2}{g'^2}x^2 - \frac{2g''}{g'}xy + Ax + By + C = 0.$$

A, B, and C being constant quantities, a comparison of the co-efficients of x^2 and xy will shew that this projection is a parabola. The projection upon one of the other planes will also give a parabola, consequently the line passed over is a parabola.

It may be proved that the curve passed over is of single curvature or upon a plane, without obtaining the equations

mon to all the bodies; it is therefore impossible to judge concerning the absolute motion of the system of which we make a part, by the appearances we observe, and it is this which characterises the law of the proportionality of the force to the velocity.

Again, it results from No. 5, that if we project each force and their resultant upon a fixed plane, the sum of the moments of the composing forces thus projected, with respect to a fixed point taken upon the plane, is equal to the moment of the projection of the resultant; but if we draw a radius, which we shall call a radius vector, from this point to the moving body, this radius projected upon the fixed plane will trace, in consequence of each force acting separately, an area equal to the product of the projection of the line which the moving body is made to describe, into one half of the perpendicular drawn from the fixed point to this projection: this area is therefore proportional to the time.

---

of projection, by extracting t from the three equations of motion, which gives one of the form
$$ax + by + cz = 0$$
that belongs to a plane surface.

If the velocities $\dfrac{dx}{dt}, \dfrac{dy}{dt},$ and $\dfrac{dz}{dt}$ parallel to the three axes at any instant whatever, are composed into one, it will be
$$v = \sqrt{\{(v' + g't)^2 + (v'' + g''t)^2 + (v''' + g'''t)^2\}}.$$
The accelerating force in the direction of the motion, or $\dfrac{dv}{dt}$, is, as may easily be proved;
$$\dfrac{g'(v' + g't) + g''(v'' + g''t) + g'''(v''' + g'''t)}{\sqrt{\{(v' + g't)^2 + (v'' + g''t)^2 + (v''' + g'''t)^2\}}}$$

It is also in a given time proportional to the moment of the projection of the force; thus, the sum of the areas which the projection of the radius vector would describe in consequence of each composing force, if it acted alone, is equal to the area which the resultant would make the same projection describe. It therefore follows, that if a body is projected in a right line, and afterwards solicited by any forces whatever directed towards a fixed point, its radius vector will always describe about this point areas proportional to the times; because the areas which the new composing quantities cause this radius to describe will be nothing. Inversely, we may see that if the moving body describes areas proportional to the times about the fixed point, the resultant of the new forces which solicit it is always directed towards this point.

7. Let us next consider the motion of a point solicited by forces, such as gravity, which seem to act continually.

The causes of this, and similar forces which have place in nature, being unknown, it is impossible to discover whether they act without interruption, or, after successive imperceptible intervals of time; but it is easy to be assured that the phænomena ought to be very nearly the same in the two hypotheses; for if we represent the velocity of a body upon which a force acts incessantly by the ordinate of a curve whose abscissa represents the time, this curve in the second hypothesis will be changed into a polygon of a very great number of sides, which for this reason may be confounded with the curve. We shall, with geometers, adopt the first hypothesis, and suppose that the interval of time which separates two consecutive actions of any force whatever is equal to the element dt of the time, which we will denote by t. It is evidently necessary to suppose that the action of the force is more considerable as the interval is greater which separates its successive actions; in order that after the same time t, the velocity may be the same; the instantaneous action of a force ought therefore to be supposed in the ratio of its intensity, and of the element of the time during which it is supposed to act. Thus, representing this intensity by P, we ought to suppose at the commencement of each instant dt, the moving body to be solicited by a force $P.dt$, and moved uniformly during this instant. This agreed upon:

It is possible to reduce all the forces which solicit a point M to three, P, Q, and R, acting parallel to three rectangular co-ordinates x, y, and z, which determine the position of this point; we shall suppose these forces to act in a contrary direction to the origin of the co-ordinates, or to tend to increase them. At the commencement of a new instant dt, the moving body receives in the direction of each of its co-ordinates, the increments of force or of velocity, $P.dt$, $Q.dt$, $R.dt$. The velocities of the point M parallel to these co-ordinates are $\frac{dx}{dt}$, $\frac{dy}{dt}$, and $\frac{dz}{dt}$; for during an indefinitely small time, they may be supposed to be uniform, and, therefore, equal to the elementary spaces divided by the element of the time. The velocities by which the moving body is actuated at the commencement of a new instant, are consequently,

$$\frac{dx}{dt} + P.dt \; ; \quad \frac{dy}{dt} + Q.dt \; ; \quad \frac{dz}{dt} + R.dt \; ;$$

or

$$\frac{dx}{dt} + d.\frac{dx}{dt} - d.\frac{dx}{dt} + P.dt \; ;$$

$$\frac{dy}{dt} + d.\frac{dy}{dt} - d.\frac{dy}{dt} + Q.dt \; ;$$

$$\frac{dz}{dt}+d.\frac{dz}{dt}-d.\frac{dz}{dt}+R.dt;$$

but at this new instant, the velocities by which the moving body is actuated parallel to the co-ordinates x, y, and z, are evidently $\frac{dx}{dt}+d.\frac{dx}{dt}$; $\frac{dy}{dt}+d.\frac{dy}{dt}$; and $\frac{dz}{dt}+d.\frac{dz}{dt}$; the forces $-d.\frac{dx}{dt}+P.dt$, $-d.\frac{dy}{dt}+Q.dt$, and $-d.\frac{dz}{dt}+R.dt$, ought therefore to be destroyed, so that the moving body may in consequence of these sole forces be in equilibrio. Thus denoting by δx, δy, and δz any variations whatever of the three co-ordinates x, y, and z, variations which it is not necessary to confound with the differentials dx, dy, and dz, that express the spaces which the moving body describes parallel to the co-ordinates during the instant dt; the equation (b) of No. 3 will become

$$0 = \delta x.\left\{d.\frac{dx}{dt}-P.dt\right\}+\delta y.\left\{d.\frac{dy}{dt}-Q.dt\right\}$$
$$+\delta z.\left\{d.\frac{dz}{dt}-R.dt\right\}. \quad (f)$$

If the point M be free, we shall equal the co-efficients of δx, δy, and δz separately to nothing, and, supposing the element dt of the time constant, the differential equations will become *

$$\frac{d^2 x}{dt^2}=P\ ;\ \frac{d^2 y}{dt^2}=Q\ ;\ \frac{d^2 z}{dt^2}=R.$$

---

* The equations $\frac{d^2 x}{dt^2}=P$, $\frac{d^2 y}{dt^2}=Q$, and $\frac{d^2 z}{dt^2}=R$, are sufficient to enable us to discover the velocity, the trajectory

If the point M be not free but subjected to move upon a curve line or a surface, there must be extracted

---

and the place at any given time, of a point not constrained to move along a line or a surface, but continually acted upon by forces which are given every instant both in magnitude and direction.

Thus supposing for greater simplicity, that the point moves in the plane xy, P and Q being constant or variable but given; by extracting the time from the two equations $\frac{d^2 x}{dt^2}=P$, $\frac{d^2 y}{dt^2}=Q$, and integrating them twice, we shall find a relation between x and y which will give the trajectory of the point. In a like manner, the relation between x and t, or y and t may be found, which will give the position of the point for any given value of the time t. The values of $\frac{dx}{dt}$ and $\frac{dy}{dt}$ will likewise give the velocities of the point in the directions of x and y, from which we may obtain the real velocity v of the point; for

$$v = \frac{ds}{dt} = \sqrt{\left\{\left(\frac{dx}{dt}\right)^2 + \left(\frac{dy}{dt}\right)^2\right\}}.$$

The first of the two constant quantities which the above double integration requires, will be determined by the value of the velocity at a given instant, such as the commencement of the time t. The second will depend upon the situation of the point with respect to the two axes at this instant.

If the moving body be attracted towards a fixed point by a single force, the integrals of the equations

$$\frac{d^2 x}{dt^2}=P, \quad \frac{d^2 y}{dt^2}=Q, \quad \frac{d^2 z}{dt^2}=R,$$

may be readily obtained in the following manner,

Let the origin A of the co-ordinates be placed at this fixed point, and suppose the moving body m in any position, having x, y, and z for its rectangular co-ordinates; then its

from the equation *(f)*, by means of the equations to the surface or the curve *, as many of the variations

---

distance from the point A will be represented by $\sqrt{x^2+y^2+z^2}$ and the force acting upon it by $-S$, which when resolved in the directions of x, y, and z gives

$$P = \frac{-Sx}{\sqrt{x^2+y^2+z^2}}, \quad Q = \frac{-Sy}{\sqrt{x^2+y^2+z^2}}, \quad \& \; R = \frac{-Sz}{\sqrt{x^2+y^2+z^2}}.$$

The three first mentioned equations, by proper multiplication and subtraction, evidently give the three following,

$$xd^2y - yd^2x = (xQ - yP).dt^2,$$
$$zd^2x - xd^2z = (zP - xR).dt^2,$$
$$yd^2z - zd^2y = (yR - zQ).dt^2.$$

If the above values of P, Q, and R are substituted in these equations, their second members will vanish, and their first will give by integration, the following $xdy - ydx = cdt$, $xdy - ydx = c'dt$, and $ydz - zdy = c''dt$, c, c', and c'' being constant quantities; these equations shew, as will be hereafter demonstrated, that equal areas are described in equal times, by the projections of the line Am upon the planes of the co-ordinates. If these integrals be added together, after having multiplied the first by z, the second by y, and the third by x, the equation $cz + c'y + c''x = 0$, which belongs to a plane, will be obtained.

* If a point moves upon a curve line or surface, it may be supposed free, and acted upon by a force equal and opposite to the perpendicular pressure upon the curve or surface. Let us, for example, suppose that $z = f(x,y)$ is an equation to a curve surface, by differentiating it, we shall have

$$dz = \left(\frac{dz}{dx}\right)dx + \left(\frac{dz}{dy}\right)dy, \text{ or, if } p = \left(\frac{dz}{dx}\right) \text{ and } q = \left(\frac{dz}{dy}\right),$$

$dz = pdx + qdy$. Let $M = \sqrt{(1+p^2+q^2)}$, then it may be easily proved that the normal of the curve surface forms with the axes x, y, and z, angles, the co-sines of which are

δx, δy, and δz as it will have equations, and the co-efficients of the remaining variations must be equalled to nothing *.

$\frac{p}{M}$, $\frac{q}{M}$, and $\frac{1}{M}$ respectively. If N represent a force in the direction of the normal, equal and opposite to the pressure upon the curve, its components in the directions of the axes x, y, and z will be $-\frac{pN}{M}$, $-\frac{qN}{M}$ and $\frac{N}{M}$ respectively; the two first forces are negative, because they tend to diminish the co-ordinates x and y, if the curve surface have its convexity towards the planes of xz and yz, as can easily be proved. The point may therefore be regarded as free, and acted upon by the forces $P-\frac{Np}{M}$, $Q-\frac{Nq}{M}$, and $R+\frac{N}{M}$; which will give the following equations,

$$\frac{d^2 x}{d t^2} = P - \frac{Np}{M},$$

$$\frac{d^2 y}{d t^2} = Q - \frac{Nq}{M},$$

$$\frac{d^2 z}{d t^2} = R + \frac{N}{M}.$$

* When the motion takes place in a resisting medium, the resistance of the medium may be regarded as a force which acts in a direction contrary to the motion of the body. Let I represent this resistance, then its moment will be $-I.\delta i$, if i is supposed to be equal to $\sqrt{(x-l)^2+(y-m)^2+(z-n)^2}$, l, m, and n being the co-ordinates of the origin of the force I. By differentiation

$$\delta i = \frac{x-l}{i}.\delta x + \frac{y-m}{i}.\delta y + \frac{z-n}{i}.\delta z.$$

If the origin of the force I is supposed to be in the tangent of the curve described by the body, and indefinitely near to

8. It is possible in the equation *(f)* to suppose the variations δx, δy, and δz equal to the differentials dx, dy, and dz, because these differentials are necessarily subjected to the conditions of the motion of the moving

---

it; it may be conceived that $x - b = dx$, $y - m = dy$, $z - n = dz$, which gives, by representing the element of the curve by ds, the following equations,

$$\frac{x-l}{i} = \frac{dx}{ds}, \quad \frac{y-m}{i} = \frac{dy}{ds}, \text{ and } \frac{z-n}{i} = \frac{dz}{ds};$$

consequently,

$$\delta i = \frac{dx}{ds}.\delta x + \frac{dy}{ds}.\delta y + \frac{dz}{ds}.\delta z.$$

If the resisting medium be in motion, it will be necessary to compose this motion with that of the body, in order to obtain the direction of the resisting force. Let $d\alpha$, $d\beta$, and $d\gamma$ denote the small spaces through which the medium passes parallel to the axes of the co-ordinates x, y, and z, during the time that the body describes the space ds, it will be proper to substract these quantities from dx, dy, and dz, in order to have the relative motions. As $ds = \sqrt{dx^2 + dy^2 + dz^2}$, if it be supposed that

$$d\sigma = \sqrt{(dx-d\alpha)^2 + (dy-d\beta)^2 + (dz-d\gamma)^2},$$

the following equation may be obtained,

$$\delta i = \frac{dx-d\alpha}{d\sigma}.\delta x + \frac{dy-d\beta}{d\sigma}.\delta y + \frac{dz-d\gamma}{d\sigma}.\delta z.$$

It should be observed with respect to the resistance I, that it is generally a function of the velocity $\frac{ds}{dt}$; but in this case in which the medium is in motion, it is a function of the relative velocity $\frac{d\sigma}{dt}$.

body M. By making this supposition, and afterwards integrating the equation (f), we shall have:

$$\frac{dx^2+dy^2+dz^2}{dt^2}=c+2\textstyle\int(P.dx+Q.dy+R.dz);$$

c being a constant quantity. $\frac{dx^2+dy^2+dz^2}{dt^2}$ is the square of the velocity of M, which velocity we will denote by v; supposing therefore that $P\,dx+Q.dy+R.dz$ is the exact differential of a function φ, we shall have *

$$v^2=c+2\varphi. \quad (g)$$

---

* In every case in which the formula $P.dx+Q.dy+R.dz$ is an exact differential of the variables x, y, and z, the equation (g) will give the velocity of the point M at any part of its trajectory, if we know it at any one determinate place. For as φ is a function of x, y, and z, let it be represented by $f(x,y,z)$, also suppose A to denote the known velocity at the point where the co-ordinates are a, a', and a'', then we shall have the equations

$$v^2=c+2f(x,y,z),$$
$$A^2=c+2f(a,a',a''),$$

and consequently

$$v^2-A^2=2f(x,y,z)-2f(a,a',a'').$$

This equation shews the value of v, when A and the co-ordinates x, y, z, a, a', and a'' corresponding to the velocities v and A are given.

It appears from the above that it is possible to determine the difference of the squares of the velocities at two points of the trajectory, by means of the co-ordinates of these points, without knowing the curve along which the moving body passes in going from one point to the other.

The above does not hold good if $P.dx+Q.dy+R.dz$ be not an exact differential, as for instance, when the forces

This case has place when the forces which solicit the point M, are functions of the respective distances of their origins from this point, which comprehends nearly all the forces of nature.

In fact S, S', &c. representing these forces, s, s', &c. being the distances of the point M from their origins, the resultant of all these forces multiplied by the variation of its direction, will be equal, by No. 2. to $\Sigma . S . \delta s$; it is also equal to $P.\delta x + Q.\delta y + R.\delta z$; we have therefore

$$P.\delta x + Q.\delta y + R.\delta z = \Sigma . S . \delta s;$$

and as the second member of this equation is an exact differential, the first is likewise. It results from the equation (g), 1st. That if the point M is not solicited by any forces, its velocity is constant, because in this case $\varphi = 0$ *. It is easy to be assured of this otherwise

---

P, Q, and R arise from friction or the resistance of a fluid, and contain in their values the velocities $\frac{dx}{dt}$, $\frac{dy}{dt}$, and $\frac{dz}{dt}$; in which case the expression is not an exact differential of a function of x, y, and z, regarded as independent variables. For to integrate φ under such circumstances, it would be necessary to substitute the values of these variables and their differentials in functions of the time, which could not be done except the problem had been previously solved.

* If the point is not acted upon by any accelerating force, but moves from an initial impulse, the forces P, Q, and R are nothing and φ vanishes, therefore $v^2 = c$, or the velocity is constant, consequently the velocities in the directions of the axes x, y, and z are constant, and may be supposed re-

by observing that a body moving on a surface or curve line, loses at each rencounter with the indefinitely small plane of the surface, or of the indefinitely small side of the curve, but an indefinitely small portion of its velocity of the second order *. 2ndly, That the point M in passing from one given point with a given velocity, to arrive at another, will have at this last point the same velocity, whatever may be the curve which it shall have described.

But if the body is not forced to move upon a determinate curve, the curve described by it possesses a si-

---

spectively equal to the invariable quantities c', c'', and c'''. We therefore have the equations

$$\frac{dx}{dt} = c', \quad \frac{dy}{dt} = c'', \quad \frac{dz}{dt} = c''';$$

from which by extracting dt, we shall obtain $c''dx = c'dy$, and $c'' dy = c' dz$; which give by integration $c''x = a + c'y$, and $c'''y = b + c''z$ for the equations of the projections of the trajectory upon the planes of xy and yz; but these equations belong to straight lines, consequently the trajectory is a straight line.

* It is proved in most treatises upon mechanics, that if a body moves along a system of inclined planes, the velocity lost in passing from one plane to another, is, as the versed sine of the angle which the planes make with each other. If the number of planes in a given curve are indefinitely increased, the supplements of their angles of inclination, and consequently the chords become indefinitely small; by the rules of trigonometry we have versed sine $= \dfrac{\text{chord}^2}{2\text{radius}}$; therefore if the chord is an indefinitely small quantity of the first order, the versed sine is an indefinitely small one of the second, consequently the velocity lost may be regarded as an indefinitely small quantity of the second order.

milar property to which we have been conducted by metaphysical considerations, and which is, in fact, but a remarkable result of the preceding differential equations. It consists in this, that the integral $\int v ds$ comprised between the two extreme points of the curve described, is less than any other curve, if the body is free, or less than any other curve subjected to the same surface upon which it moves, if it is not entirely free.

To make this appear, we shall observe that $P.dx + Q.dy + R.dz$ being supposed an exact differential, the equation (g) gives

$$v \delta v = P.\delta x + Q.\delta y + R.\delta z;$$

the equation (f) of the preceding number becomes also

$$0 = \delta x . d.\frac{dx}{dt} + \delta y . d.\frac{dy}{dt} + \delta z . d.\frac{dz}{dt} - vdt . \delta v.$$

Naming the element of the curve described by the moving body ds, we shall have

$$vdt = ds; \qquad ds = \sqrt{dx^2 + dy^2 + dz^2};\ *$$

and by equating,

$$0 = \delta x . d.\frac{dx}{dt} + \delta y . d.\frac{dy}{dt} + z.d.\frac{dz}{dt} - ds.\delta v; \qquad (h)$$

by differentiating, with respect to δ, the expression of ds, we shall have

$$\frac{ds}{dt}.\delta.ds = \frac{dx}{dt}.\delta.dx + \frac{dy}{dt}.\delta.dy + \frac{dz}{dt}.\delta.dz.$$

---

* That $ds = \sqrt{(dx^2 + dy^2 + dz^2)}$ is evident from considering that the co-ordinates of s and $s + ds$ are x, y, z and $x + dx, y + dy, z + dz$; consequently,
$$s + ds - s = \sqrt{\{(x + dx - x)^2 + (y + dy - y)^2 + (z + dz - z)^2\}},$$
which gives $ds = \sqrt{(dx^2 + dy^2 + dz^2)}$.

The characteristics d and δ being independant*, we may place them one before the other at will; the preceding equation can therefore be made to take the following form,

$$v.\delta ds = \frac{d.\{dx.\delta x + dy.\delta y + dz.\delta z\}}{dt} - \delta x.d.\frac{dx}{dt} - \delta y.d.\frac{dy}{dt}$$
$$- \delta z.d.\frac{dz}{dt};$$

---

* It may here be necessary to observe, that when equations contain the differentials dx, dy, and dz, and the variations δx, δy, and δz at the same time, the differentials and variations are to be supposed constant with respect to each other, in all the various processes of differentiation or integration. The order in which these processes are performed is also indifferent as to the result. Thus $\delta.dx = d.\delta x$, $\delta.d^2x = d.\delta.dx = d^2\delta x$, $\delta.d^n x = d^m.\delta.d^{n-m}x = d^n\delta x$, also $\delta \int u = \int \delta u$, u being here supposed a function of x, y, z, dx, dy, dz, d^2x, &c., the sign $\int$ denoting the integration of the function with respect to the characteristics dx, &c. If u be a function of x, y, and z, the equation $u = f(x,y,z)$ gives

$$du = \left(\frac{du}{dx}\right)dx + \left(\frac{du}{dy}\right)dy + \left(\frac{du}{dz}\right)dz,$$

also

$$\delta u = \left(\frac{\delta u}{\delta x}\right)\delta x + \left(\frac{\delta u}{\delta y}\right)\delta y + \left(\frac{\delta u}{\delta z}\right)\delta z.$$

in which we evidently have from the process of differentiation

$$\left(\frac{du}{dx}\right) = \left(\frac{\delta u}{\delta x}\right), \quad \left(\frac{du}{dy}\right) = \left(\frac{\delta u}{\delta y}\right), \quad \left(\frac{du}{dz}\right) = \left(\frac{\delta u}{\delta z}\right).$$

As the nature of these notes will not permit me to enter fully upon the subject of variations, I shall refer the reader, who is desirous of information respecting them, to the Traité du Calcul Differentiel et Integral pour S. F. Lacroix. The Traité Elementaire de Calcul Differentiel et de Calcul Integral, by the same author, contains an abridged account of them from the large work.

by subtracting from the first member of this equation, the second member of the equation (h), we shall have

$$\delta.(vds) = \frac{d.(dx.\delta x + dy.\delta y + dz.\delta z)}{dt}.$$

This last equation being integrated by relation to the characteristic d, will give

$$\delta.\int vds = \text{const.} + \frac{dx.\delta x + dy.\delta y + dz.\delta z}{dt}.$$

If we extend the integral to the entire curve described by the moving body, and if we suppose the extreme points of this curve to be invariable *, we shall have $\delta.\int v d s = 0$; that is to say, of all the curves along which a moving body, subjected to the forces P, Q, and R, can pass from one given point to another given point, it will describe that in which the variation of the integral $\int vds$ is nothing, and in which, consequently this integral is a minimum. If the point moves along a curve surface without being acted upon by any force, its velocity is constant, and the integral $\int vds$ becomes

---

* If the point from which the body begins to move be fixed, the quantities δx, δy, and δz are there respectively equal to nothing, therefore the constant quantity of the equation

$$\delta\int vds = \text{const.} + \frac{dx.\delta x + dy.\delta y + dz.\delta z}{dt}$$

is equal to nothing, as its other terms vanish at that point. If the quantities δx, δy, and δz are also respectively equal to nothing at the end of the motion, from the point where it ends being fixed, we shall have $\frac{dx.\delta x + dy.\delta y + dz.\delta z}{dt}$ equal to nothing, therefore $\delta\int vds$ is equal to nothing, that is, the variation of the quantity $\int vds$ is a minimum.

vds; thus the curve described by the moving body is, in this case, the shortest which it is possible to trace upon the surface, from the point of departure to that of arrival *.

---

* Maupertuis, in two memoirs, one sent to the Academy of Sciences at Paris, in the year 1744, and the other to that of Berlin, in the year 1746, asserted, that in all the changes which take place in the situation of a body, the product of the mass of the body by its velocity and the space which it has passed over is a minimum. This he called the principle of the least action, and it was applied by him to the discovery of the laws of the refraction and the reflection of light, the laws of the collision of bodies, the laws of equilibrium, &c. Euler afterwards shewed that in the trajectories of bodies acted upon by central forces, the integral of the velocity multiplied by the element of the curve is always a minimum, which is an excellent application of the principle of the least action to the motions of the planets. This general principle, which was assumed as a metaphysical truth, appears evidently to be derived from the laws of mechanics.

The following will be sufficient to shew the reader the way in which the principle may be used to discover the laws of the refraction and of the reflection of light.

Suppose a ray of light to pass from one point to another, if the points are in the same medium, the velocity of the ray is constant, the path is a straight line and the principle obvious; if they are in two different mediums, let v represent the velocity of the ray, and s the space it passes through, in the first medium, and v' its velocity and s' the space passed through by it in the second medium; we shall then have the quantity $vs+v's'$, which is a minimum for the value of $\int vds$. The solution of this question is very easy and leads us to the following equation, $v.\sin. a = v.\sin. b$, in which a represents the angle of incidence and b the angle of refraction at

9. Let us determine the pressure which a point moving upon a surface exerts against it. Instead of extracting from the equation *(f)* of No. 7, one of the variations δx, δy, and δz, by means of the equation of the surface, we may by No. 3, add to this equation the differential equation of the surface multiplied by an indeterminate quantity $-\lambda dt$, and afterwards consider the three variations δx, δy, and δz as independant quantities. Let therefore $u=0$ be the equation of the surface; we shall add to the equation *(f)* the term $-\lambda.\delta u.dt$, and the pressure with which the point acts against it will be, by No. 3, equal to

$$\lambda.\sqrt{\left(\frac{du}{dx}\right)^2+\left(\frac{du}{dy}\right)^2+\left(\frac{du}{dz}\right)^2}.$$

Let us now suppose that the point is not solicited by any force, its velocity v will be constant; if we observe lastly, that $vdt=ds$, the element dt of the time being supposed constant, the element ds of the curve described will be so likewise, and the equation *(f)* aug-

---

the surface of the second medium. The above equation shews that the ratio of the two sines depends upon that of the velocities of the ray in passing through the different mediums.

If the ray in passing from one point to another is reflected at the surface of the second medium, the velocity will be constant, and the path a minimum; in which case, it may be readily proved, that the angle of incidence of the ray in passing from one point to the surface of the second medium, is equal to the angle of reflection from it to the other point.

mented by the term $-\lambda.\delta x\, dt$, will give the three following,

$$0 = v^2 \cdot \frac{d^2x}{dt^2} - \lambda.\left(\frac{du}{dx}\right);\ 0 = v^2 \cdot \frac{d^2y}{dt^2} - \lambda.\left(\frac{du}{dy}\right);$$

$$0 = v^2 \cdot \frac{d^2z}{dt^2} - \lambda.\left(\frac{du}{dz}\right);$$

from which we may obtain

$$\lambda \sqrt{\left(\frac{du}{dx}\right)^2 + \left(\frac{du}{dy}\right)^2 + \left(\frac{du}{dz}\right)^2} = \frac{v^2 \sqrt{(d^2x)^2+(d^2y)^2+(d^2z)^2}}{ds^2};$$

but as ds is constant, the radius of curvature of the curve described by the moving body is equal to

$$\frac{ds^2}{\sqrt{(d^2x)^2+(d^2y)^2+(d^2z)^2}};$$

by naming this radius r, we shall have

$$\lambda = \sqrt{\left(\frac{du}{dx}\right)^2 + \left(\frac{du}{dy}\right)^2 + \left(\frac{du}{dz}\right)^2} = \frac{v^2}{r};$$

that is to say, the pressure exercised by the point against the surface, is equal to the square of its velocity divided by the radius of curvature of the curve which it describes.

If the point move upon a spherical surface, it will describe the circumference of a great circle of the sphere which passes by the primitive direction of its motion: for there is not any reason why it should move more to the right than to the left of the plane of this circle; its pressure against the surface, or, what comes to the same, against the circumference which it describes, is therefore equal to the square of its velocity divided by the radius of this circumference.

If we imagine the point to be attached to the end of a thread supposed without thickness, having the other extremity fastened to the centre of the surface; it is evident that the pressure exercised by this point against the circumference will be equal to the tension which the thread would experience if the point were retained by it alone. The effort which this point makes to stretch the thread, and to go farther from the centre of the circumference, is, what is called the centrifugal force; therefore the centrifugal force is equal to the square of the velocity divided by the radius.

In the motion of a point upon any curve whatever, the centrifugal force is equal to the square of the velocity, divided by the radius of curvature of the curve, because the indefinitely small arc of this curve is confounded with the circumference of the circle of curvature; we shall therefore have the pressure which the point exerts against the curve that it describes, by adding to the square of the velocity divided by the radius of curvature, the pressure due to the forces which solicit this point. In the motion of a point upon a surface, the pressure due to the centrifugal force, is equal to the square of the velocity divided by the radius of curvature of the curve described by this point, and multiplied by the sine of the inclination of the plane of the circle of curvature to the tangential plane of the surface * : by adding to this pressure that which

---

* Suppose the radius of curvature RP or r, (*fig.* 7.), of the point P of the curve described by the body upon the surface, to be produced to A; let PA represent the centrifugal force $\dfrac{V^2}{r}$ of the body moving in the curve; from P draw the line PB perpendicular to the plane tangent of the surface at

arises from the action of the forces that solicit the point, we shall have the whole pressure which it exerts against the surface.

We have seen that if the point is not acted upon by any forces, its pressure against the surface is equal to the square of its velocity divided by the radius of cur-

---

P; draw AB perpendicular to PB, let the line DPC be the section of the plane tangent caused by a plane passing through the points ABP perpendicular to the plane tangent; then as AB, CPD, are respectively perpendicular to the line PB, they are parallel to each other, therefore the angle BAP is equal to the angle DPR, but this last angle is that which the plane of the circle of curvature makes with the plane tangent, for as the intersection of the plane of the curve and of the plane tangent of the surface is the tangent to the curve at P, the line PB is perpendicular to it; likewise the radius of curvature of the curve is perpendicular to its tangent at the same point, consequently the plane passing through the lines BP, PR, and the line PD in it drawn from the point P, are perpendicular to the tangent of the curve; therefore the angle DPR is the angle which the plane tangent makes with the plane of the curve. By trigonometry in the right angled triangle PAB, we have

$$\text{rad.} (1) : \text{sine } BAP \text{ or } DPR :: PA\left(\frac{V^2}{r}\right) : PB,$$

therefore $PB = \dfrac{V^2}{r}$. sine DPR, but PB represents the centrifugal force of the body moving on the surface, consequently the centrifugal force of a body moving upon a surface, is equal to the square of the velocity divided by the radius of curvature of the curve described by this point, and multiplied by the sine of the inclination of the plane of the circle of curvature to the tangential plane of the surface.

vature of the curve described, the plane of the circle of curvature, that is to say, the plane which passes by two consecutive sides of the curve described by the point, is, in this case, perpendicular to the surface. This curve relative to the surface of the earth, is called a perpendicular to the meridian, and we have proved (No. 8) that it is the shortest which it is possible to draw from one point to another upon the surface.

10. Of all the forces which we observe upon the earth, the most remarkable is gravity; it penetrates into the most inward parts of bodies, and without the resistance of the air, would make them fall with an equal velocity. Gravity is very nearly the same at the greatest heights to which we are able to ascend, and at the lowest depths to which we are able to descend; its direction is perpendicular to the horizon; but in the motions of projectiles, we may suppose, without sensible error, that it is constant, and that it acts along parallel lines; on account of the small extent of the curves which they describe relative to the surface of the earth. These bodies moving in a resisting fluid, we shall call β the resistance that they experience, it is directed along the side of the curve described by them, which side we will denote by ds, we shall moreover call g the force of gravity. This agreed upon:

Let us resume the equation (f) of No. 7, and suppose the plane of x and y horizontal, and the origin of z at the most elevated point; the force β will produce in the directions of x, y, and z, the three forces $-\beta.\frac{dx}{ds}$, $-\beta.\frac{dy}{ds}$, and $-\beta.\frac{dz}{ds}$; we shall therefore have, by No. 7,

$$P = -\beta.\frac{dx}{ds}; \quad Q = -\beta.\frac{dy}{ds}; \quad R = -\beta.\frac{dz}{ds} + g;$$

and the equation (f) will become

$$0 = \delta x \cdot \left\{ d \cdot \frac{dx}{dt} + \beta \cdot \frac{dx}{ds} \cdot dt \right\} + \delta y \cdot \left\{ d \cdot \frac{dy}{dt} + \beta \cdot \frac{dy}{ds} \cdot dt \right\}$$
$$+ \delta z \cdot \left\{ d \cdot \frac{dz}{dt} + \beta \cdot \frac{dz}{ds} \cdot dt - g\, dt \right\}.$$

If the body be entirely free, we shall have the three equations

$$0 = d \cdot \frac{dx}{dt} + \beta \cdot \frac{dx}{ds} \cdot dt \,; \quad 0 = d \cdot \frac{dy}{dt} + \beta \cdot \frac{dy}{ds} \cdot dt \,;$$
$$0 = d \cdot \frac{dz}{dt} + \beta \cdot \frac{dz}{ds} \cdot dt - g\, dt.$$

The two first will give

$$\frac{dy}{dt} \cdot d \cdot \frac{dx}{dt} - \frac{dx}{dt} \cdot d \cdot \frac{dy}{dt} = 0\,;$$

from which, by integration, we shall obtain $dx = f\, dy$; f being a constant quantity. This is the equation to an horizontal right line; therefore the body moves in a vertical plane. By taking for this plane that of x and z, we shall have $y = 0$; the two equations

$$0 = d \cdot \frac{dx}{dt} + \beta \cdot \frac{dx}{ds} \cdot dt \,; \quad 0 = d \cdot \frac{dz}{dt} + \beta \cdot \frac{dz}{ds} \cdot dt - g\, dt\,;$$

will give, by making dx constant,

$$\beta = \frac{ds \cdot d^2 t}{dt^3}\,; \quad 0 = \frac{d^2 z}{dt} - \frac{dz \cdot d^2 t}{dt^2} + \beta \cdot \frac{dz}{ds} \cdot dt - g\, dt.$$

from which we may obtain $g\, dt^2 = d^2 z$, and by differentiating $2g\, dt \cdot d^2 t = d^3 z$; by substituting for $d^2 t$ its value $\frac{\beta\, dt^3}{ds}$, and for dt^2 its value $\frac{d^2 z}{g}$ we shall have

$$\frac{\beta}{g} = \frac{ds \cdot d^3 z}{2 \cdot (d^2 z)^2}$$

This equation gives the law of the resistance β necessary to make a projectile describe a determinate curve. If the resistance be proportional to the square of the velocity, β is equal to $h \cdot \frac{ds^2}{dt^2}$, h being constant in the case where the density of the medium is uniform. We shall then have

$$\frac{\beta}{g} = \frac{h\,ds^2}{g\,dt^2} = \frac{h.ds^2}{d^2z};$$

and from substituting the value of $\frac{\beta}{g}$, $hds = \frac{d^3z}{2\,d^2z}$; which gives by integration

$$\frac{d^2z}{dx^2} = 2a..c^{2hs};$$

a being a constant quantity*, and c the number whose

---

* The value of a may be found as follows: Let gdt^2 be substituted for d^2z in the equation $\frac{d^2z}{dx^2} = 2ac^{2hs}$, then it is evident that $\frac{dx^2}{dt^2} = \frac{g}{2a}.c^{2bs}$, but $\frac{dx}{dt}$ is the velocity of the projectile parallel to the axis of x, let v represent the velocity of projection, and θ the angle which its direction makes with that axis, then $\left(\frac{dx}{dt}\right)^2 = v^2\cos.^2\theta$; consequently as $s=0$ at the commencement of the motion, we have $v^2\cos.^2\theta = \frac{g}{2a}$ and $a = \frac{g}{2v^2.\cos.^2\theta}$,

therefore $\frac{d^2z}{dt^2} = \frac{g}{v^2\cos.^2\theta}.c^{2hs}$.

By supposing dz equal to pdx, this last equation may be changed into the following $\frac{dp}{dx} = \frac{g}{v^2.\cos.^2\theta}.c^{2hs}$; but as $ds = \sqrt{dx^2 + dz^2} = dx\sqrt{1+p^2}$, by multiplying the first member of this equation by $dx\sqrt{1+p^2}$ and the second by ds it will become $dp\sqrt{1+p^2} = \frac{g}{v^2\cos.^2\theta}.c^{2hs}\,ds$. The integral of $dp\sqrt{1+p^2}$ is $\frac{1}{2}\{p\sqrt{1+p^2} + \log.(p+\sqrt{1+p^2})\}$, and that of

hyperbolical logarithm is unity. If we suppose the resistance of the medium to be nothing, or $h=0$, we shall have, by integration, the equation to a parabola

$$z=ax^2+bx+e\;;$$

b and e being constant quantities.

$\frac{g}{v^2.\cos.^2\theta}.c^{2hs}\,ds$ is $\frac{g}{v^2.\cos.^2\theta}.\frac{c^{2hs}}{2h}$, therefore $p\sqrt{1+p^2}$

$+\log.(p+\sqrt{1+p^2})=C+\frac{g}{v^2.\cos.^2\theta}.\frac{c^{2hs}}{h}$, C being a constant quantity, which may be determined by observing, that at the commencement of the projection $s=0$ and $p=\tang.\theta$, consequently, $C=\tang.\theta\sqrt{1+\tang.^2\theta}+\log.(\tang.\theta+\sqrt{1+\tan.^2\theta})$

$-\frac{g}{v.^2\cos.^2\theta.h}$. By substituting in the above integral the value of c^{2hs} obtained from the equation $\frac{dp}{dx}=\frac{g}{v.^2\cos.^2\theta}.c^{2hs}$, we shall have the following

$$dx=\frac{dp}{h\{p\sqrt{1+p^2}+\log.(p+\sqrt{1+p^2})-C\}};$$

which gives, as $dz=pdx$,

$$dz=\frac{pdp}{h\{p\sqrt{1+p^2}+\log.(p+\sqrt{1+p^2})-C\}}.$$

By the integration of these equations the values of x and z would be given in functions of p. A third equation may be obtained which would give the time in a function of the same quantity, by substituting the value of dx derived from the equation $dt^2=\frac{dp.dx}{g}$ in one of the preceding equations, and extracting the square root, which will give the following

$$dt=\frac{dp}{\sqrt{gh\{p\sqrt{1+p^2}+\log.(p+\sqrt{1+p^2})-C\}^{\frac{1}{2}}}}.$$

If these three equations could be integrated so as to have a

The differential equation $d\dot{x} = g dt^2$ will give $dt^2 = \frac{2a}{g} . dx^2$, from which we may obtain $t = x\sqrt{\frac{2a}{g}} + f'$. If x, z, and t are supposed to begin together, we shall

---

finite form, the complete solution of the problem would be obtained. In this case, the two first, by the elimination of p, would give the trajectory of the curve. This problem has exercised the skill of many eminent mathematicians from the time of Newton to that of Legendre, but all their solutions are very complicated. The trajectory may be described from points by means of the two first equations, and tables made for forming it at any inclination. Vide the memoir of Moreau, in the eleventh cahier of the Journal de l' Ecole Polytechnique. The descending branch of the curve has an asymtote, as appears from making p indefinitely great, which gives $dx = \frac{dp}{hp^2}$ and $dz = \frac{dp}{hp}$, or by integration, $x = c' - \frac{1}{hp}$ and $z = c'' + \frac{1}{h} \log . p$, c' and c'' being constant quantities.

From these two equations it appears, that if p be indefinitely increased, the value of z will become indefinitely great, although it does not increase so fast as p, and that of x will approximate to c' as its limit. If, therefore, on the horizontal axis of x, at a distance equal to c', from the origin of the co-ordinates, a perpendicular be let fall, that line will be an asymtote to the descending branch of the curve.

When the angle of projection of the body is very small with respect to the horizon, and the initial velocity not considerable, that part of the curve above the horizontal line of projection may be readily found by approximation, and is applicable to the case of ricochet firing. Vide a memoir of Borda amongst those of the Academie des Sciences, 1769.

have $e=0$ and $f'=0$; consequently $t=x\sqrt{\frac{2a}{g}}$, and $z=ax^2+bx$, which give

$$z=\frac{gt^2}{2}+bt^2\cdot\sqrt{\frac{g}{2a}}.$$

These three equations contain the whole theory of projectiles in a vacuum; it results from the above, that the velocity is uniform in an horizontal direction, and in a vertical one, it is the same as that which would be acquired by the body falling down the vertical.

If the body falls from a state of rest, b will vanish, and we shall have $\frac{dz}{dt}=gt$; $z=\frac{1}{2}gt^2$; the velocity therefore increases as the time, and the space increases as the square of the time.

It is easy, by means of these formulæ, to compare the centrifugal force to that of gravity. It has been shewn by what precedes, that v being the velocity of a body moving in the circumference of a circle, whose radius is r, the centrifugal force is $\frac{v^2}{r}$. Let h be the height from which it ought to fall to acquire the velocity v; we shall have by what precedes, $v^2=2gh$; from which we may obtain $\frac{v^2}{r}=g\cdot\frac{2h}{r}$. If $h=\frac{1}{2}r$, the centrifugal force becomes equal to the gravity g; thus a heavy body attached to the extremity of a thread fastened by its other extremity to an horizontal plane, will stretch this thread with the same force as if it were suspended vertically, provided, that it moves upon this plane with the velocity which it would have acquired by falling from a height equal to half the length of the thread,

11. Let us consider the motion of a heavy body on a spherical surface.

By naming its radius r, and fixing the origin of the co-ordinates x, y, and z at its centre; we shall have $r^2-x^2-y^2-z^2=0$; this equation compared with that of $u=0$, gives $u=r^2-x^2-y^2-z^2$: by adding therefore to the equation (f) of No. 7, the function δu multiplied by the indeterminate $-\lambda dt$, we shall have

$$0=\delta x\left\{d.\frac{dx}{dt}+2\lambda x.dt\right\}+\delta y\left\{d.\frac{dy}{dt}+2\lambda y.dt\right\}+\delta z\left\{d.\frac{dz}{dt}+2\lambda z.dt-g dt\right\};$$

an equation in which we may equal separately to nothing, the co-efficients of each of the variations δx, δy, and δz, which will give the three following equations,

$$\left.\begin{array}{l}0=d.\dfrac{dx}{dt}+2\lambda x.dt;\\[4pt]0=d.\dfrac{dy}{dt}+2\lambda y.dt;\\[4pt]0=d.\dfrac{dz}{dt}+2\lambda z.dt-g dt\end{array}\right\}.\quad(A)$$

The indeterminate λ makes known the pressure which the moving body exerts against the surface. This pressure is by No. 9 equal to $\lambda\sqrt{\left(\dfrac{du}{dx}\right)^2+\left(\dfrac{du}{dy}\right)^2+\left(\dfrac{du}{dz}\right)^2}$; it is consequently equal to $2\lambda r$; but by No. 8, we have

$$c+2gz=\frac{dx^2+dy^2+dz^2}{dt^2},$$

c being a constant quantity; by adding this equation to the equations (A) divided by dt, and multiplied respectively by x, y, and z; and observing, lastly, that the differential equation of the surface is $0=xdx+ydy+zdz$, which, by differentiation, gives

$$0=xd^2x+yd^2y+zd^2z+dx^2+dy^2+dz^2;$$

we shall find
$$2\lambda r = \frac{c + 3gz}{r}.$$

If we multiply the first of the equations (A) by $-y$, and add it to the second multiplied by x, we shall have from integrating their sum
$$\frac{xdy - ydx}{dt} = c',$$
c' being a new constant quantity.

The motion of a point is thus reduced to three differential equations of the first order
$$xdx + ydy = -zdz\ ;$$
$$xdy - ydx = c'dt\ ;$$
$$\frac{dx^2 + dy^2 + dz^2}{dt^2} = c + 2gz.$$

By raising each member of the two first equations to the square, and then adding them together, we shall have
$$(x^2 + y^2).(dx^2 + dy^2) = c'^2 dt^2 + z^2 dz^2\ ;$$
if we substitute in the place of $x^2 + y^2$ its value $r^2 - z^2$, and in the place of $\frac{dx^2 + dy^2}{dt^2}$ its value $c + 2gz - \frac{dz^2}{dt^2}$ we shall have, by supposing that the body departs from the vertical,
$$dt = \frac{-rdz}{\sqrt{(r^2 - z^2).(c + 2gz) - c'^2}}.$$

The function under the root may be changed to the following form, $(a - z).(z - b).(2gz + f)$; a, b, and f being determined by the equations
$$f = \frac{2g(r^2 + ab)}{a + b}\ ;$$
$$0 = \frac{2g(r^2 - a^2 - ab - b^2)}{a + b}\ ;$$
$$c'^2 = \frac{2g.(r^2 - a^2).(r^2 - b^2)}{a + b}$$

It is possible thus to substitute for the constant quantities c and c' the new ones a and b; the first of which is the greatest value of z, and the second the least. By making afterwards

$$\sin \theta = \sqrt{\frac{a-z}{a-b}},$$

the preceding differential equation will become

$$dt = \frac{r.\sqrt{2.(a+b)}}{\sqrt{g.\{(a+b)^2+r^2-b^2\}}} \cdot \frac{d\theta}{\sqrt{1-\gamma^2.\sin^2\theta}};$$

γ^2 being equal to $\dfrac{a^2-b^2}{(a+b)^2+r^2-b^2}$.

The angle θ gives the co-ordinate z by means of the equation

$$z = a.\cos^2\theta + b.\sin^2\theta,$$

and the co-ordinate z divided by r, gives the co-sine of the angle which the radius r makes with the vertical.

Let ϖ be the angle which the vertical plane passing by the radius r, makes with the vertical plane passing by the axis of x; we shall then have *

$$x = \sqrt{r^2-z^2}.\cos\varpi; \quad y = \sqrt{r^2-z^2}.\sin\varpi;$$

which give $xdy - ydx = (r^2-z^2).d\varpi$; the equation $xdy - ydx = c'dt$ will also give

$$d\varpi = \frac{c'dt}{r^2-z^2},$$

---

* For $\sqrt{r^2-z^2}$ is the projection of the line r upon the plane of xy, and if from the extremity of $\sqrt{r^2-z^2}$ a perpendicular be drawn to the axis of x, we shall have $\sqrt{r^2-z^2}$: x :: rad. (1) : cos. ϖ, therefore $x = \sqrt{r^2-z^2}.\cos\varpi$. In a similar manner we shall have $y = \sqrt{r^2-z^2}.\sin\varpi$.

by substituting for z and dt their preceding values in θ, we shall have the angle ϖ in a function of θ; thus we may know at any time whatever the two angles θ and ϖ; which is sufficient to determine the position of the moving body.

Naming the time which is employed in passing from the highest to the lowest value of z, the semi-oscillation of the body; let $\frac{1}{2}T$ represent this time. To determine it, it is necessary to integrate the preceding value of dt from $\theta = 0$ to $\theta = \frac{1}{2}\pi$; π being the semi-circumference of a circle whose radius is unity: we shall thus find

$$T = \pi \sqrt{\frac{r}{g}} \cdot \sqrt{\frac{2r.(a+b)}{(a+b)^2 + r^2 - b^2}} \cdot \left\{ 1 + \left(\frac{1}{2}\right)^2 \cdot \gamma^2 + \left(\frac{1.3}{2.4}\right)^2 \cdot \gamma^4 + \left(\frac{1.3.5}{2.4.6}\right)^2 \cdot \gamma^6 + \&c. \right\}.$$

Supposing the point to be suspended from the extremity of a thread without mass, which is fixed at its other extremity, if the length of the thread is r the point will move exactly as in the interior of a spherical surface, and it will form with the thread a pendulum, the co-sine of whose greatest distance from the vertical will be $\frac{b}{r}$. If we suppose that in this state, the velocity of the moving body is nothing, it will oscillate in a vertical plane, and in this case we shall have $a = r$, $\gamma^2 = \frac{r-b}{2r}$. The fraction $\frac{r-b}{2r}$ is the square of the sine of half the greatest angle which the thread forms with the vertical; the entire duration T of the oscillation of the pendulum will therefore be

$$T = \pi\sqrt{\frac{r}{g}} \cdot \left\{1 + \left(\frac{1}{2}\right)^2 \cdot \left(\frac{r-b}{2r}\right) + \left(\frac{1.3}{2.4}\right)^2 \cdot \left(\frac{r-b}{2r}\right)^2 \right.$$
$$\left. + \left(\frac{1.3.5}{2.4.6}\right)^2 \cdot \left(\frac{r-b}{2r}\right)^3 + \&c. \right\}.$$

If the oscillation is very small, $\frac{r-b}{2r}$ is a very small fraction, which may be neglected, and we shall have

$$T = \pi \cdot \sqrt{\frac{r}{g}};$$

the very small oscillations are therefore isochronous or of the same duration, whatever may be their extent; and we can easily, by means of this duration, and of the corresponding length of the pendulum, determine the variations of the intensity of gravity at different parts of the earth's surface.

Let z be the height from which gravity makes a body fall during the time T, we shall have, by No. 10, $2z = gT^2$, and consequently $z = \frac{1}{2}\pi^2 \cdot r$; we shall therefore have with very great precision by means of the length of a pendulum that beats seconds, the space through which gravity will cause bodies to fall during the first second of their descent. From experiments very exactly made, it appears, that the length of a pendulum vibrating seconds is the same, whatever may be the substances which are made to oscillate: from which it results that gravity acts equally upon all bodies, and that it tends in the same place, to impress upon them the same velocity in the same time.

12. The isochronism of the oscillations of a pendulum being only an approximation, it is interesting to know the curve upon which an heavy body ought to move, to arrive at the same time upon the point where its motion ceases, whatever may be the arc which it shall de-

scribe from the lowest point. But to solve this problem in the most general manner, we will suppose, conformably to what has place in nature, that the point moves in a resisting medium. Let s represent the arc described from the lowest point to the curve, z the vertical abscissa reckoned from this point; dt the element of the time, and g the gravity. The retarding force along the arc of the curve will be, first, the gravity resolved along the arc ds, which becomes equal to $g \cdot \frac{dz}{ds}$; secondly, the resistance of the medium, which we shall express by $\varphi \cdot \left(\frac{ds}{dt}\right)$, $\left(\frac{ds}{dt}\right)$ being the velocity of the moving body, and $\varphi \cdot \left(\frac{ds}{dt}\right)$ being any function whatever of this velocity. The differential of this velocity will be, by No. 7, equal to $-g \cdot \frac{dz}{ds} - \varphi \left(\frac{ds}{dt}\right)$; we shall therefore have by making dt constant

$$0 = \frac{d^2 s}{dt^2} + g \cdot \frac{dz}{ds} + \varphi \cdot \left(\frac{ds}{dt}\right). \quad (i)$$

Let us suppose $\varphi \cdot \left(\frac{ds}{dt}\right) = m \cdot \frac{ds}{dt} + n \cdot \frac{ds^2}{dt^2}$, and $s = \psi(s')$; if we denote by $\psi'(s')$ the differential of $\psi(s')$ divided by ds', and by $\psi''(s')$ that of $\psi'(s')$ divided by ds', we shall have

$$\frac{ds}{dt} = \frac{ds'}{dt} \cdot \psi'(s'),$$

$$\frac{d^2 s}{dt^2} = \frac{d^2 s'}{dt^2} \cdot \psi'(s') + \frac{ds'^2}{dt^2} \cdot \psi''(s');$$

and the equation (i) will become

$$0 = \frac{d^2 s'}{dt^2} + m \cdot \frac{ds'}{dt} + \frac{ds'^2}{dt^2} \left\{ \frac{\psi''(s') + n\{\psi'(s')\}^2}{\psi'(s')} \right\} +$$

$$\frac{g \cdot dz}{ds'\{\psi'(s')\}^2}; \quad (t)$$

we shall cause the term multiplied by $\frac{ds'^2}{dt^2}$ to disappear by means of the equation
$$0 = \psi''(s') + n.[\psi'(s')]^2; \ast$$
this equation gives by integration
$$\psi(s') = \log.\{h.(s'+q)^{\frac{1}{n}}\} = s,$$
h and q being constant quantities. If we make s' to commence with s, we shall have $hq^{\frac{1}{n}} = 1$, and if, for greater simplicity, we make $h = 1$, we shall have
$$s' = c^{ns} - 1;$$
c being the number, the hyperbolical logarithm of which is unity: the differential equation (l) then becomes
$$0 = \frac{d^2 s'}{dt^2} + m.\frac{ds'}{dt} + n^2 g.\frac{dz}{ds'}.(1+s')^2$$

---

* The integral of $0 = \psi''(s') + n.[\psi'(s')]^2$ may be readily found, by substituting for $\psi''(s')$, and $\psi'(s')$ their values; the equation then becomes $0 = \dfrac{d^2 s}{ds} \dfrac{d^2 s'}{ds'} + nds$, that by integration gives $h.\log.ds - h.\log.ds' + ns = e$, or $h.\log.\dfrac{ds}{ds'} = e - ns$, from which we may obtain $ds' = \dfrac{c^{ns}}{c^e}.ds$, c being the number whose hyperbolical logarithm is unity; the integral of this equation is $s' + q = \dfrac{c^{ns} - e}{n}$, which gives $h.\log.(n.(s'+q)) = ns - e$, or $h.\log.\{n.(s'+q)\}^{\frac{1}{n}} + \dfrac{e}{n} = s$; if $\dfrac{e}{n} = h$. $\log.\dfrac{h}{n^{\frac{1}{n}}}$, then hyp. log. $\{h(s'+q)^{\frac{1}{n}}\} = s$.

By supposing s' very small, we may develope the last term of this equation in an ascending series, with respect to the powers of s', which will be of this form, $ks' + ls'^i + \&c.$, i being greater than unity; the last equation then becomes

$$0 = \frac{d^2 s'}{dt'^2} + m \cdot \frac{ds'}{dt} + ks' + ls'^i + \&c.$$

This equation being multiplied by

$$c^{\frac{mt}{2}} \cdot \{\cos. \gamma t + \sqrt{-1} \cdot \sin. \gamma t\},$$

and afterwards integrated, supposing γ equal to $\sqrt{k - \frac{m^2}{4}}$, will be changed into

$$c^{\frac{mt}{2}} \cdot \{\cos. \gamma t + \sqrt{-1} \cdot \sin. \gamma t\} \cdot \left\{\frac{ds'}{dt} + \left(\frac{m}{2} - \gamma \cdot \sqrt{-1}\right) \cdot s'\right\} = -l \cdot \int s'^i dt \cdot c^{\frac{mt}{2}} \{\cos. \gamma t + \sqrt{-1} \cdot \sin. \gamma t\} - \&c.$$

By comparing separately, the real and the imaginary parts, we shall have two equations, by means of which $\frac{ds'}{dt}$ may be extracted; but here it will be sufficient for us to consider the following

$$c^{\frac{mt}{2}} \cdot \frac{ds'}{dt} \cdot \sin. \gamma t + c^{\frac{mt}{2}} \cdot s' \cdot \left\{\frac{m}{2} \cdot \sin. \gamma t - \gamma \cdot \cos. \gamma t\right\} = -l \cdot \int s'^i dt \cdot c^{\frac{mt}{2}} \cdot \sin. \gamma t - \&c.;$$

the integrals of the second member being supposed to commence with t. Naming T the value of t at the end of the motion, when $\frac{ds}{dt}$ is nothing; we shall have at that instant

$$c^{\frac{mT}{2}} \cdot s' \cdot \left\{\frac{m}{2} \cdot \sin. \gamma T - \gamma \cdot \cos. \gamma T\right\} = -l \cdot \int s'^i \cdot dt \cdot c^{\frac{mt}{2}} \cdot \sin. \gamma t - \&c.$$

In the case of s' being indefinitely small, the second member of the equation will be reduced to nothing when compared with the first, and we shall have

$$0 = \frac{m}{2}.\sin.\gamma T - \gamma.\cos.\gamma T;$$

from which we may obtain

$$\tang.\gamma T = \frac{2\gamma}{m};$$

and as the time T is, by the supposition, independent of the arc passed over, this value of the $\tang.\gamma T$ has place for any arc whatever, which will give for any value of s'

$$0 = l.\int s'^i.dt. \; c^{\frac{mt}{s}}.\sin.\gamma t + \&c.$$

the integral being taken from $t = 0$, to $t = T$.

By supposing s' very small, this equation will be reduced to its first term, and it can only be satisfied by making $l = 0$; for the factor $c^{\frac{mt}{s}}.\sin.\gamma t$ being always positive from $t = 0$ to $t = T$, the preceding integral is necessarily positive in this interval. It is not therefore possible to have tautochronism but on the supposition of

$$n^2 g.\frac{dz}{ds'}.(1+s')^2 = ks';$$

which gives for the equation of the tautochronous curve

$$g dz = \frac{k\, ds}{n}(1 - c^{-ns}).$$

In a vacuum, and when the resistance is proportional to the simple velocity, n is nothing, and this equation becomes $gdz = ksds$; which is the equation to the cycloid*.

It is remarkable that the co-efficient n of the part of the resistance proportional to the square of the velocity, does not enter into the expression of the time T; and it is evident by the preceding analysis, that this expression will be the same, if we add to the preceding law of the resistance, the terms

$$p \cdot \frac{ds^3}{dt^3} + q \cdot \frac{ds^4}{dt^4} + \&c.$$

If in general, R represents the retarding force along the curve, we shall have

$$0 = \frac{d^2s}{dt^2} + R.$$

s is a function of the time t, and of the whole arc passed

---

* The cycloid is the only curve in a plane that is tautochronous in a vacuum, but this property belongs to an indefinite number of curves of double curvature, which may be formed by applying a cycloid to a vertical cylinder of any base, without changing the altitudes of the points of the curve above the horizontal plane. This is evident from considering the equation $v^2 = c + 2\varphi$ of No. 8, which by proper substitution becomes $\frac{ds^2}{dt^2} = c - 2gz$ and gives $dt = \pm \frac{ds}{\sqrt{c - 2gz}}$; the upper sign being taken if t and s increase together, and the lower, if one increases whilst the other decreases. From this last equation it appears, that the value of t depends upon the initial velocity and the relation between the vertical ordinates and the arcs of the curve If therefore this velocity and this relation be the same, when the curve is changed, the above equation will not be altered any more than the law of motion which it denotes.

over, which is consequently a function of t and of s. By differentiating this last function, we shall have a differential equation of the form

$$\frac{ds}{dt} = V;$$

V being a function of t and s, which by the condition of the problem ought to be nothing, when t has a value which is indeterminate and independent of the arc passed over. Suppose for example $V = S.T$, S being a function of s alone, and T a function of t alone; we shall have

$$\frac{d^2 s}{dt^2} = T \cdot \frac{dS}{ds} \cdot \frac{ds}{dt} + S \cdot \frac{dT}{dt} = \frac{dS}{Sds} \cdot \frac{ds^2}{dt^2} + S \cdot \frac{dT}{dt};$$

but the equation $\frac{ds}{dt} = ST$, gives t, and consequently $\frac{dT}{dt}$ equal to a function of $\frac{ds}{S.dt}$; which function we will denote by $\frac{ds^2}{S^2.dt^2} \cdot \psi\left(\frac{ds}{Sdt}\right)$; we shall therefore have

$$\frac{d^2 s}{dt^2} = \frac{ds^2}{S.dt^2} \cdot \left\{ \frac{dS}{ds} + \psi\left(\frac{ds}{Sdt}\right) \right\} = -R.$$

Such is the expression of the resistance which answers to the differential equation $\frac{ds}{dt} = ST$; and it is easy to perceive that it comprises the case of the resistance proportional to the two first powers of the velocity, multiplied respectively by constant co-efficients. Other differential equations would give different laws of resistance.

CHAP. III.

Of the equilibrium of a system of bodies.

13. The most simple case of the equilibrium of many bodies, is that of two material points which strike each other with equal and directly contrary velocities; their mutual impenetrability evidently destroys their velocities, and reduces them to a state of rest. Let us now consider a number m of contiguous material points disposed in a right line, and actuated by the velocity u, in the direction of this line. Let us suppose, in like manner, a number m' of contiguous points, disposed upon the same right line, and actuated by a velocity u' directly contrary to u, so that the two systems shall strike each other. In order that they may be in equilibrio at the moment of the impact, there ought to be a relation between u and u' which it is necessary to determine.

For this purpose we shall observe that the system m, actuated by the velocity u, will reduce a single material point to a state of equilibrium, if it be actuated by a velocity mu in a contrary direction; for each point in the system will destroy a velocity equal to u in this last point, and consequently its m points will destroy the entire velocity mu: we may therefore substitute for this system a single point actuated by the velocity mu.

We can, in like manner, substitute for the system m' a single point actuated by the velocity $m'u'$; but the two systems being supposed to cause equilibrium, the two points which take their places ought in like manner to do it, this requires that their velocities should be equal; we have therefore for the condition of the equilibrium of the two systems $mu = m'u'$.

The mass of a body is the number of its material points, and the product of the mass by its velocity is called its quantity of motion; this is what is understood by the force of a body in motion.

For the equilibrium of two bodies or of two systems of points which strike each other in contrary directions, the quantities of motion, or the opposite forces, ought to be equal, and consequently the velocities should be inversely as the masses.

The density of bodies depends upon the number of material points which are contained in a given volume. In order to have their absolute density, it would be necessary to compare their masses with that of a body without pores; but as we know no bodies of that description, we can only speak of the relative densities of bodies; that is to say, the ratio of their density to that of a given substance. It is evident that the mass is in the ratio of the magnitude and the density, by naming M the mass of the body, U its magnitude, and D its density, we shall have generally $M = DU$: an equation in which it should be observed that the quantities M, D, and U express a certain relation to the unities of their species.

What we have said, is on the supposition that bodies are composed of similar material points, and that they differ only by the respective positions of these points. But as the nature of bodies is unknown, this hypothesis is at least precarious; and it is possible that there may be essential differences between their ultimate particles. Happily the truth of this hypothesis is of no consequence to the science of mechanics, and we may make use of it without fearing any error, provided that by similar material points, we understand points which by striking each other with equal and opposite velocities, mutually produce equilibrium whatever may be their nature.

14. Two material points of which the masses are m and m', cannot act upon each other but along the line that joins them. In fact, if the two points are connected by a thread which passes over a fixed pulley, their reciprocal action cannot be directed along this line. But the fixed pulley may be considered as having at its centre a mass of infinite density, which re-acts upon the two bodies m and m', whose action upon each other may be considered as indirect.

Let p denote the action which m exercises upon m', by the means of a straight line inflexible and without mass, which is supposed to unite the two points. Conceive this line to be actuated by two equal and opposite forces p and $-p$, the force $-p$ will destroy in the body m a force equal to p, and the force p of the right line will be communicated entirely to the body m'. This loss of force in m, occasioned by its action upon m', is what is called the re-action of m'; thus in the communication of motions, the re-action is always equal and contrary to the action. From observation it appears that this principle has place in all the forces of nature.

Let us suppose two heavy bodies m and m' attached to the extremities of an horizontal right line inflexible and without mass, which can turn freely about one of its points.

To conceive the action of the bodies upon each other when they produce the state of equilibrium, it is necessary to suppose an indefinitely small bend in the right line at its fixed point; we shall then have two right lines making at this point an angle, which differs from two right angles by an indefinitely small quantity ω. Let f and f' represent the distances m and m' from this fixed point: by resolving the weight of m into two forces, one acting upon the fixed point, and the other directed towards m': this last force will be represented by $\frac{mg \cdot (f+f')}{\omega f'}$, g being the force of gravity*. The

---

* Let mCm' (fig. 8,) represent the lever, which is supposed to be bent from the horizontal right line mCn at C, so that the angle $m'Cn = \omega$, then $mC = f, m'C = f'$, the line $m'n$ perpendicular to the line mC continued $= \omega f'$ nearly, and the line mm' joining the bodies m and $m' = f+f'$ nearly. Complete the rectangular parallelogram $mnm'o$, and suppose mo or its equal nm' to represent the weight of m; this force may be resolved into two others, represented in quantity and direction by the lines mm' and $m'o$; we shall then have $mo = nm'$ $(\omega f')$ to mm' $(f+f')$, as mg (the weight of m) is to $\frac{mg(f+f')}{\omega f'}$ or the force with which m acts upon the body m'. The force with which m' acts upon m may be found in a similar manner.

action of m' upon m will in like manner be $\frac{m\,g(f+f')}{\omega f}$;
by equalling these two forces, in consequence of their equilibrium, we shall have $mf = m'f'$; this gives the known law of the equilibrium of the lever, and at the same time enables us to conceive the reciprocal action of parallel forces.

Let us now consider the equilibrium of a system of points m, m', m'', &c. solicited by any forces whatever, and re-acting upon each other. Let f be the distance of m from m'; f' the distance of m from m''; and f'' the distance of m' from m'', &c.; again, let p be the reciprocal action of m upon m'; p' that of m upon m'': p'' that of m' upon m'' &c.; lastly, let $mS, m'S', m''S''$, &c. be the forces which solicit m, m', m'', &c.; and s, s', s'', &c. the right lines drawn from their origins unto the bodies m, m', m'', &c.

This being agreed upon, the point m may be considered as perfectly free and in equilibrio, in consequence of the force mS, and the forces which the bodies m, m', m'', &c. communicate to it: but if it were subjected to move upon a surface or a curve, it would be necessary to add the re-action of the surface or of the curve to these forces. Let δs be the variation of s; and let $\delta_{i}f$ denote the variation of f taken by regarding m' as fixed. Denoting in like manner, by $\delta_{i}f'$, the variation of f', taken by regarding m'' as fixed, &c. Let R and R' represent the re-actions of two surfaces, which form by their intersection the curve upon which the point m is forced to move, and δr, $\delta r'$ the variations of the directions of these last forces. The equation (d) of No. 3, will give

$0 = m.S\delta s + p.\delta_{i}f + p'.\delta_{i}f' + \&c. + R.\delta r + R'.\delta r'$

L

In like manner, m' may be considered as a point which is perfectly free and in equilibrio, in consequence of the force $m'S'$, of the actions of the bodies m, m'', &c., and of the re-actions of the surfaces upon which it is obliged to move; which re-actions we shall denote by R'' and R'''. Let $\delta s'$ be therefore the variation of s'; $\delta_{u} f$ the variation of f, taken by regarding m as fixed; $\delta_{,}f''$ the variation of f'', taken by regarding m'' as fixed, &c. Moreover let $\delta r''$ and $\delta r'''$ be the variations of the directions of R'' and R'''; the equilibrium of m' will give

$$0 = m'S'.\delta s' + p.\delta_u f + p''.\delta_{,}f'' + \&c. . + R''.\delta r'' + R'''\delta r'''.$$

If we form similar equations relative to the equilibrium of m'', m''', &c.; by adding them together and observing that *

$$\delta f = \delta_{,} f + \delta_{u} f\,; \quad \delta f' = \delta_{,} f' + \delta_{u} f'\,; \&c.$$

---

* The ninth diagram will serve to render this more evident. Suppose that the line joining the bodies m and m' is represented by f; that the point m' being immoveable, the point m passes over the indefinitely small space mn. Join nm'; from the point n let fall the perpendicular na upon the line mm', then ma will represent the projection of the line mn upon the line mm', see notes No. 2, and we shall have $\overline{ma} = \overline{mm'}(f) - \overline{am'}$, but $\overline{am'}^{2} = \overline{m'n}^{2} - \overline{na}^{2}$, and as $\overline{na}^{2}$ is an indefinitely small quantity of the second order, it may be neglected, therefore $\overline{am'} = \overline{nm'}$ nearly, consequently $\overline{ma} = \overline{mm'}(f) - \overline{nm'} = \delta_{,}f$. Again, let m' be supposed to pass over the indefinitely short space $m'n'$, whilst m remains immoveable, join mn', then $\overline{mm'} - \overline{mn'} = \delta_{u} f$. If m and m' be

δf, $\delta f'$, &c. being the whole variations of f, f', &c. we shall have

$$0 = \Sigma . m . S . \delta s + \Sigma . p . \delta f + \Sigma . R . \delta r, \quad (k)$$

an equation in which the variations of the co-ordinates of the different bodies of the system are entirely arbitrary. It should here be observed that instead of $m . S . \delta s$, it is possible in consequence of the equation (a) of No. 2, to substitute the sum of the products of all the partial forces by which m is actuated, multiplied by the variations of their respective directions. It is the same with the products $m' S' . \delta s'$, $m'' S'' . \delta s''$, &c.

If the bodies m, m', m'', &c. are invariably connected with each other, the distances f, f', f'', &c. will be constant, and we shall have for the condition of the connection of the parts of the system, $\delta f = 0, \delta f' = 0, \delta f'' = 0$, &c. The variations of the co-ordinates in the equation (k) being arbitrary, we may subject them to satisfy these last equations, and then the forces p, p', p'', &c. which depend upon the reciprocal action of the bodies of the system, will disappear from this equation: we can also cause the terms $R . \delta r$, $R' . \delta r'$, &c. to disappear, by subjecting the variations of the co-ordinates to satisfy the equations of the surfaces upon which the bodies are forced to move, the equation (k) thus becomes

$$0 = \Sigma . m . S . \delta s ; \quad (l)$$

from which it follows, that in the case of equilibrium, the sum of the variations of the products of the forces

---

supposed to vary at the same time, and move respectively to n and n', let n and n' be joined, then we shall have $mm' - nn' = \delta f = \delta_{,} f + \delta_{,,} f$, by neglecting indefinitely small quantities of higher orders than the first.

by the elements of their directions is nothing, in whatever manner we make the position of the system to vary, provided that the conditions of the connection of its parts be observed.

This theorem which we have obtained on the particular supposition of a system of bodies invariably connected together, is general, whatever may be the conditions of the connection of the parts of the system. To demonstrate this, it is sufficient to shew, that by subjecting the variations of the co-ordinates to these conditions, we shall have in the equation (k)

$$0 = \Sigma . p . \delta f + \Sigma . R . \delta r;$$

but it is evident that $\delta r, \delta r'$, &c. are nothing, in consequence of these conditions; it is therefore only required to prove that we have $0 = \Sigma . p . \delta f$, by subjecting the variations of the co-ordinates to the same conditions.

Let us imagine the system to be acted upon by the sole forces p, p', p'', &c. and let us suppose that the bodies are obliged to move upon the curves which they would describe in consequence of the same conditions. Then these forces may be resolved into others, one part q, q', q'', &c. directed along the lines f, f', f'', &c. which would mutually destroy each other, without producing any action upon the curves described; another part T, T', T'', &c. perpendiculars to the curves described, lastly, the remaining part tangents to these curves, in consequence of which the system will be moved*. But

---

* In (fig. 10) where only two bodies m and m' are considered, mm' is the line joining the bodies, AmB the curve upon which m is forced to remain, mp the force p that acts upon m in the direction of the line mm' or f, rp or q that

it is easy to perceive that these last forces ought to be nothing; for the system being supposed to obey them freely, they are not able to produce either pressure upon the curves described, or re-action of the bodies upon each other; they cannot therefore make equilibrium to the forces $-p, -p', -p''$, &c. q, q', q'', &c. T, T', T'', &c.; it is consequently necessary that they should be nothing, and that the system should be in equilibrio by means of the sole forces $-p, -p', -p''$, &c. q, q', q'', &c. T, T', T'', &c. Let $\delta i, \delta i'$, &c. represent the variations of the directions of the forces T, T', &c.; we shall then have in consequence of the equation (k)

$$0 = \Sigma.(q-p).\delta f + \Sigma.T.\delta i;$$

but the system being supposed to be in equilibrio by means of the sole forces q, q', &c. without any action resulting upon the curves described; the equation (k) again gives $0 = \Sigma.q.\delta f$, which reduces the above to the following

$$0 = \Sigma.p.\delta f - \Sigma.T.\delta i.$$

If we subject the variations of the co-ordinates to answer to the equations of the described curves, we shall have $\delta i = 0$, $\delta i' = 0$, &c. and the above equation becomes

$$0 = \Sigma.p.\delta f;$$

as the curves described are themselves arbitrary, and

---

part of it which is destroyed by the mutual action of the the bodies without producing any effect upon the curve AmB, mr the remaining force of p, which is resolved into the force mT, or T that acts perpendicularly to the curve, and Tr which acts in the direction of a tangent to it.

only subjected to the conditions of the connection of the parts of the system; the preceding equation has place, provided that these conditions be fulfilled, and then the equation (*k*) will be changed into the equation (*l*). This equation is the analytical traduction of the following principle, known under the name of the principle of virtual velocities.

"If we make an indefinitely small variation in the position of a system of bodies, which are subjected to the conditions that it ought to fulfil: the sum of the forces which solicit it, each multiplied by the space that the body to which it is applied moves along its direction, should be equal to nothing in the case of the equilibrium of the system."*

---

* The following are proofs of the truth of the principle of virtual velocities in the cases of the lever, the inclined plane, and the wedge.

First, with respect to the lever; let mCm' (*fig.* 11.) represent a straight lever in equilibrio upon the fulcrum C, by means of the forces S and S' acting in the respective directions of the lines $mS=s$, and $m'S'=s'$; if this lever be supposed to be disturbed in an indefinitely small degree, so that m and m' describe the arcs mn and $m'n'$ respectively, and perpendiculars na and $n'b$ be drawn from n and n', upon the directions of the forces S and S', we shall have $ma = -\delta s$, and $m'b = \delta s'$. Let the perpendiculars Cc and Cd be drawn to the directions of the forces S and S' from the fulcrum C, then as the indefinitely small arcs mn and $m'n'$ may be supposed rectilinear, and the angles Cmn and $Cm'n'$ right angles, we shall have ang. $amn=$ ang. mCc, and ang. dCm' $=$ ang. $n'm'b$; and consequently the rectangular triangles

This equation has not only place in the case of equilibrium, but it assures its existence. Suppose that the

---

amn, mCc, and dCm', $n'm'b$ respectively similar, therefore by proportion
$$ma : mn :: Cc : Cm,$$
also
$$m'b : m'n' :: Cd : Cm',$$
which give $ma = \dfrac{mn}{Cm}.Cc$, and $m'b = \dfrac{m'n'}{Cm'}.Cd$, consequently $\delta s = -\dfrac{mn}{Cm}.Cc$, and $\delta s' = \dfrac{m'n'}{Cm'}.Cd$. Let these values of δs and $\delta s'$ be substituted in the equation of virtual velocities
$$S.\delta s + S'.\delta s' = 0$$
observing that $\dfrac{mn}{Cm} = \dfrac{m'n'}{Cm'}$, and we shall find that $S.Cc = S'.Cd$; which is a well-known property of the lever.

In the case of the inclined plane, let the twelfth diagram be supposed to represent a section formed by a plane passing through the centres of gravity m and m', of two weights in equilibrio upon two inclined planes AB and BC, which have the common altitude BD, the weights being connected by a string passing over the pulley P. Let the position of the weights be changed so that their centres of gravity m and m' may pass through the indefinitely small spaces mn and $m'n'$ respectively; from m and m' let the lines $mS = s$, and $m'S' = s'$ be drawn in the direction of gravity, and suppose the weight of the body resting upon AB to be represented by S and that of the body resting upon BC by S': from n and n' draw the perpendiculars na and $n'b$ respectively to the lines mS and $m'S'$, or their continuations. Then as the lines forming the triangle nma are respectively parallel to the lines forming the triangle ABD, the triangles are similar:

equation *(l)* having place, the points m, m' &c. take the velocities v, v', &c. in consequence of the forces

---

for a like reason the triangles $m'n'b$ and BCD are similar: we have therefore
$$AB : BD :: mn : ma,$$
$$BC : BD :: m'n' : m'b,$$
consequently $ma = \dfrac{BD}{AB}.mn$ and $m'b = \dfrac{BD}{BC}.m'n'$. If in the equation of virtual velocities
$$S.\delta s + S'.\delta s' = 0,$$
as $\delta s = ma$ and $\delta s' = -m'b$, their respective values $\dfrac{BD}{AB}.mn$ and $-\dfrac{BD}{BC}.m'n'$ are substituted, we shall, as $mn = m'n'$, easily obtain the following equation $S.BC = S'.AB$, which shews that the weights have the same ratio to each other as the lengths of the planes upon which they rest have; this is well known from other principles.

Lastly, in the case of the wedge, let ABC *(fig. 13)*, represent the section of a wedge, and the plane MN upon which it rests in equilibrio, from the perpendicular pressures of the forces S and S' upon its sides AB and AC, in the directions $mS = s$ and $m'S' = s'$ respectively. Suppose that in consequence of an indefinitely small variation in the situation of the wedge, it takes the position abc, its sides meeting the directions of the forces or their continuations in n and n', it is evident that the small right lines mn and $m'n'$ will be the spaces passed over by the powers S and S', in their respective directions. Join Aa, and let CA and ba be prolonged until they meet at F, and from the points A and a let the perpendiculars AH and Ag be drawn to the prolongations, then we shall evidently have $m'n' = Ga$, and $mn = AH$,

mS, $m'S'$, &c. which are applied to them. The system will be in equilibrio by means of these forces, and the forces $-mv$, $-m'v'$, &c. denoting by δv, $\delta v'$, &c. the variations of the directions of the new forces, we shall have from the principle of virtual velocities
$$0 = \Sigma . mS . \delta s - \Sigma . m . v \delta v;$$
but we have by the supposition $0 = \Sigma . mS . \delta s$; we shall therefore have $0 = \Sigma . m . v \delta v$. The variations δv, $\delta v'$, &c. ought to be subjected to the conditions of the system, they may therefore be supposed equal to vdt, $v'dt$, &c. and we shall have $0 = \Sigma . mv^2$, which equation gives $v = 0$, $v' = 0$, &c.; therefore the system is in equilibrio in consequence of the sole forces mS, $m'S'$ &c.

As the lines aF and aA are respectively parallel to the lines AB and BC, the triangles ABC and FaA are similar, consequently
$$AB : AC :: Fa : FA;$$
we have likewise, as the right angled triangles FaG and FAH, by having a common angle at F, are similar,
$$aG : AH :: Fa : FA;$$
therefore
$$AB : AC :: aG : AH.$$
This last equation, as $AH = mn = -\delta s$ and $Ga = m'n' = \delta s'$, gives $\delta s = \dfrac{AC}{AB} . -\delta s'$; If this value of δs be substituted in the equation of virtual velocities
$$S . \delta s + S' . \delta s' = 0$$
we shall obtain the following, $S . AC = S' . AB$, which shews that when the wedge is in equilibrio, the powers acting upon

The conditions of the connection of the parts of the system, may at all times be reduced to certain equations between the co-ordinates of its different bodies. Let $u=0$, $u'=0$, $u''=0$, &c. be these different equations, we shall be enabled by No. 3, to add to the equation *(l)* the function $\lambda.\delta u + \lambda'.\delta u' + \lambda''.\delta u'' + $ &c., or $S.\lambda.\delta u$; λ, λ', λ'', &c. being indeterminate functions of the co-ordinates of the bodies; the equation will then become

$$0 = \Sigma.m.S.\delta s + \Sigma.\lambda.\delta u \ ;$$

in this case the variations of all the co-ordinates are arbitrary, and we may equal their co-efficients to nothing; which will give so many equations, by means of which we can determine the functions λ, λ' &c. If we lastly compare this equation with the equation *(k)* we shall have

$$\Sigma.\lambda.\delta u = \Sigma.p.\delta f + \Sigma.R.\delta r \ ;$$

from which it will be easy to find the reciprocal actions of the bodies m, m', &c. and the pressures $-R$, $-R'$, &c. that they exercise against the surfaces on which they are forced to remain *.

---

it are to each other as the sides of the wedge to which they are applied, which is a well known property of it.

The principle of virtual velocities may readily be proved in the cases of the wheel and axle, the pulley, the screw, &c. and holds good in every case of machinery in equilibrio.

* The following examples, extracted from the Mechanique Analytique of Lagrange, will serve to shew the facility with which the principle of virtual velocities may be applied to the solution of various problems.

LAPLACE'S MECHANICS. 83

15. If all the bodies of a system are firmly attached together, its position may be determined by those of

---

It may here be premised, that the forces which act upon any point or body will be supposed to be reduced to three, P, Q, and R, acting respectively in the directions of the co-ordinates x, y, and z, and tending to diminish them. The quantities belonging to the different bodies will be distinguished by one, two, three &c. marks according to the order in which they are considered. Thus the sum of the moments of the forces which act upon the bodies will be

$$P.\delta x + Q.\delta y + R.\delta z + P'.\delta x' + Q'.\delta y' + R'.\delta z' + P''.\delta x'' + \&c.$$

To this must be added, the differentials of the equations of condition, each multiplied by an indeterminate quantity.

Let us first consider the problem of three bodies firmly attached to an inextensible thread. In this case, the conditions of the problem are, that the distances between the first and second, and between the second and third bodies, will be invariable; these distances being the lengths of the portions of the thread intercepted between them. Let f be the first of these distances, and g the second, we shall then have $\delta f = 0$, and $\delta g = 0$, for the equations of condition, therefore $\delta u = \delta f$ and $\delta u' = \delta g$, and the general equation of equilibrium will become

$$P.\delta x + Q.\delta y + R.\delta z + P'.\delta x' + Q'.\delta y' + R'.\delta z' + P''.\delta x'' + Q''.\delta y'' + R''.\delta z'' + \lambda.\delta u + \lambda'.\delta u' = 0.$$

The values of f and g are

$$f = \sqrt{\{(x'-x)^2 + (y'-y)^2 + (z'-z)^2\}},$$
$$g = \sqrt{\{(x''-x')^2 + (y''-y')^2 + (z''-z')^2\}},$$

therefore

$$\delta f = \frac{(x'-x)(\delta x' - \delta x) + (y'-y)(\delta y' - \delta y) + (z'-z)(\delta z' - \delta z)}{f}$$

three of its points which are not in the same right line; the position of each of its points depends upon three

---

$$\delta g = \frac{(x'-x)(\delta x'' - \delta x') + (y'-y)(\delta y'' - \delta y') + (z'-z)(\delta z'' - \delta z')}{g}$$

these values being substituted will give the nine following equations for the conditions of the equilibrium of the thread,

$$P - \lambda \cdot \frac{x'-x}{u} = 0,$$

$$Q - \lambda \cdot \frac{y'-y}{u} = 0,$$

$$R - \lambda \cdot \frac{z'-z}{u} = 0,$$

$$P' + \lambda \cdot \frac{x'-x}{u} - \lambda' \cdot \frac{x''-x'}{u'} = 0$$

$$Q' + \lambda \cdot \frac{y'-y}{u} - \lambda' \cdot \frac{y''-y'}{u'} = 0,$$

$$R' + \lambda \cdot \frac{z'-z}{u} - \lambda' \cdot \frac{z''-z'}{u'} = 0,$$

$$P'' + \lambda' \cdot \frac{x''-x'}{u'} = 0,$$

$$Q'' + \lambda' \cdot \frac{y''-y'}{u'} = 0,$$

$$R'' + \lambda' \cdot \frac{z''-z'}{u'} = 0.$$

It now remains to eliminate the two indeterminate quantities λ and λ' from these equations, which may be done vari-

co-ordinates, this produces nine indeterminate quantities; but the mutual distances of three points being

---

ous ways, each of which will give equations, either different, or presented differently, for the equilibrium of the three bodies.

It is evident that if we add the three first equations to the three next, and to the three last, we shall obtain the three following equations delivered from the unknown quantities λ and λ'.

$$P+P'+P''=0,$$
$$Q+Q'+Q''=0,$$
$$R+R'+R''=0,$$

which shew, that the sum of all the forces parallel to each of the three axes of x, y, and z, should be nothing.

There now remains four more equations which it is necessary to discover; for this purpose, if the three middle equations are respectively added to the three last, the three following will be obtained, which do not contain λ',

$$P'+P''+\frac{\lambda}{u}\cdot(x'-x)=0,$$
$$Q'+Q''+\frac{\lambda}{u}\cdot(y'-y)=0,$$
$$R'+R''+\frac{\lambda}{u}\cdot(z'-z)=0;$$

by the extraction of λ, the two following will be obtained.

$$Q'+Q''-\frac{y'-y}{x'-x}(P'+P'')=0,$$

given and invariable, we are enabled by their means to reduce these indeterminates to six others, which

$$R' + R'' - \frac{z'-z}{x'-x}(P' + P'') = 0.$$

Lastly, considering the three final equations, by extracting λ' from them, we shall have the two following,

$$Q'' - \frac{y''-y'}{x''-x'} \cdot P'' = 0,$$

$$R'' - \frac{z''-z'}{x''-x'} \cdot P'' = 0.$$

These seven equations contain the conditions necessary for the equilibrium of three bodies, and when joined to the given equations of condition u and u', will be sufficient for determining the position of each of them in space.

If an inextensible thread be charged with four bodies, acted upon respectively by the forces P, Q, R ; P', Q', R'; P'', Q'', R'', &c. in the directions of the three axes of $x, y,$ and z; we shall find by similar proceedings, the nine following equations for the equilibrium of these four bodies,

$$P + P' + P'' + P''' = 0,$$

$$Q + Q' + Q'' + Q''' = 0,$$

$$R + R' + R'' + R''' = 0,$$

$$Q' + Q'' + Q''' - \frac{y'-y}{x'-x}(P' + P'' + P''') = 0,$$

substituted in the equation *(l)*, will introduce six arbitrary variations; by equalling their co-efficients to

$$R' + R'' + R''' - \frac{z'-z}{x'-x}(P'+P''+P''') = 0,$$

$$Q'' + Q''' - \frac{y''-y'}{x''-x'}(P''+P''') = 0,$$

$$R'' + R''' - \frac{z''-z'}{x''-x'}(P''+P''') = 0,$$

$$Q''' - \frac{y'''-y''}{x'''-x''} \cdot P''' = 0,$$

$$R''' - \frac{z'''-z''}{x'''-x''} \cdot P''' = 0.$$

It would be easy to extend this solution to any number of bodies, or to the case of the funicular or catenarian curve.

The solution would have been in some respects simplified, if the invariability of the distances f g, &c. had been directly introduced into the calculation.

Thus, confining ourselves to the case of three bodies, and denoting by ψ and ψ' the angles which the lines f and g make with the plane of x and y; and by φ and φ' the angles which the projections of the same lines upon the same plane make with the axis of x, we shall have

$x'-x = f.\cos.\varphi.\cos.\psi;\ y'-y = f.\sin.\varphi.\cos.\psi:\ z'-z = f.\sin.\psi;$
$x''-x' = g.\cos.\varphi'.\cos.\psi';\ y''-y' = g.\sin.\varphi'.\cos.\psi';\ z''-z' = g.\sin.\psi'.$

Substituting the values of x', y', z', x'', y'', and z'' obtained

nothing, we shall have six equations that will contain all the conditions of the equilibrium of the system; let us proceed to develope these equations.

---

from these equations, in the general formula of the equilibrium of three bodies

$$P.\delta x + Q.\delta y + R.\delta z + P'.\delta x' + Q'.\delta y' + R'.\delta z' + P''.\delta x'' + Q''.\delta y'' + R''.\delta z'' = 0$$

and simply causing the quantities $x, y, z, \varphi, \varphi', \psi, \psi'$, whose variations will remain indeterminate, to vary, and equalling separately to nothing the quantities multiplied by each of these variations, we shall have the seven equations

$$P + P' + P'' = 0,$$
$$Q + Q' + Q'' = 0,$$
$$R + R' + R'' = 0,$$
$$(P' + P'')\sin.\varphi - (Q' + Q'')\cos.\varphi = 0,$$
$$P''\sin.\varphi' - Q''\cos.\varphi' = 0,$$
$$(P' + P'')\cos.\varphi.\sin.\psi + (Q' + Q'')\sin.\varphi.\sin.\psi - (R' + R'')\cos.\psi = 0,$$
$$P''\cos.\varphi'.\sin.\psi' + Q''\sin.\varphi'.\sin.\psi' - R''\cos.\psi' = 0,$$

of which the five first coincide with those found before in the question of three bodies connected by an inextensible thread, by the elimination of the indeterminate quantities λ and λ'; and the two last are readily reduced by eliminating Q' and Q'', by means of the fourth and fifth equations.

But if by this means we have more readily obtained the final equations, it is because we have employed a preliminary tranformation of the variables which contains the equations

For this purpose, if x, y, z be the co-ordinates of m; x', y', z' those of m'; x'', y'', z'' those of m'', &c. we shall have

---

of condition, instead of immediately employing the equations with indeterminate co-efficients as before, so that the equation is reduced to a pure mechanism of calculation. Moreover, we have by these co-efficients the value of the forces which the rods f and g ought to sustain from their resistance to extension, as will be shewn hereafter.

If the first body is supposed to be fixed, the differentials δx, δy, and δz vanish, and the terms affected by these differentials, will disappear of themselves from the general equation of equilibrium. In this case, the three first equations

$$P - \lambda \cdot \frac{x' - x}{u} = 0, \quad Q - \lambda \cdot \frac{y' - y}{u} = 0, \text{ and } R - \lambda \cdot \frac{z' - z}{u} = 0, \text{ will}$$

not have place, therefore the equations, $P + P' + P'' + \&c. = 0$, $Q + Q' + Q'' + \&c. = 0$, $R + R' + R'' + \&c. = 0$, will not have place, but all the others will remain the same. In this case, the thread is supposed to be fixed at one of its extremities.

If the two ends of the thread be fixed, we shall have not only $\delta x = 0$, $\delta y = 0$, $\delta z = 0$, but also $\delta x''' \&c. = 0$, $\delta y''' \&c. = 0$, $\delta z''' \&c. = 0$; and the terms affected by these six differentials, in the general equation of equilibrium, will consequently disappear, as well as the six particular equations which depend upon them.

In general, if the two extremities of a thread are not entirely free, but attached to points which move after a given law; this law expressed analytically, will give one or more equations between the differentials δx, δy, and δz, which relate to the first body, and the differentials $\delta x''' \&c.$, $\delta y''' \&c.$, $\delta z''' \&c.$, which relate to the last; and it will be necessary to add these equations, each multiplied by a new indeterminate

$$f = \sqrt{(x'-x)^2 + (y'-y)^2 + (z'-z)^2};$$
$$f' = \sqrt{(x''-x)^2 + (y''-y)^2 + (z''-z)^2};$$
$$f'' = \sqrt{(x'''-x)^2 + (y'''-y)^2 + (z'''-z)^2};$$
&c.

---

co-efficient, to the general equation of equilibrium found above; or to substitute in the general equation, the value of one or more of these differentials obtained from the above equations, and lastly to equal to nothing, the co-efficient of each that remains. As this is not attended with any difficulty, we shall omit it.

In order to discover the forces which arise from the re-action of the thread upon the different bodies, we will, in the present case, consider the equations

$$\delta u = \delta f = \frac{(x'-x)(\delta x' - \delta x) + (y'-y)(\delta y' - \delta y) + (z'-z)(\delta z' - \delta z)}{f}$$

$$\delta u = \delta g = \frac{(x''-x')(\delta x'' - \delta x') + (y''-y')(\delta y'' - \delta y') + (z''-z')(\delta z'' - \delta z')}{g}$$

&c.

With respect to the first body whose co-ordinates are x, y, and z, $\dfrac{\delta u}{\delta x} = -\dfrac{x'-x}{f}$, $\dfrac{\delta u}{\delta y} = -\dfrac{y'-y}{f}$, and $\dfrac{\delta u}{\delta z} = -\dfrac{z'-z}{f}$;

we shall therefore have

$$\sqrt{\left\{\left(\frac{\delta u}{\delta x}\right)^2 + \left(\frac{\delta u}{\delta y}\right)^2 + \left(\frac{\delta u}{\delta z}\right)^2\right\}} = \frac{\sqrt{\{(x'-x)^2 + (y'-y)^2 + (z'-z)^2}}{f} = 1.$$

Therefore the first body will be acted upon by the other bodies with a force λ, the direction of which is perpendicular to the surface represented by the equation $\delta u = \delta f = 0$, supposing the quantities x, y, and z to vary; but it is evident that this surface is that of a sphere, having f for its radius, and x', y', and z' for the co-ordinates of its centre,

If we suppose
$$\delta x = \delta x' = \delta x'' = \&c.;$$
$$\delta y = \delta y' = \delta y'' = \&c.;$$
$$\delta z = \delta z' = \delta z'' = \&c.;$$

---

consequently the force λ will be directed along this same radius, that is, along the thread which joins the first and second bodies.

With respect to the second body whose co-ordinates are x', y', z', we have
$$\frac{\delta u}{\delta x'} = \frac{x'-x}{\delta f}, \quad \frac{\delta u}{\delta y'} = \frac{y'-y}{f}, \quad \frac{\delta u}{\delta z'} = \frac{z'-z}{f};$$
therefore
$$\sqrt{\left\{\left(\frac{\delta u}{\delta x'}\right)^2 + \left(\frac{\delta u}{\delta y'}\right)^2 + \left(\frac{\delta u}{\delta z'}\right)^2\right\}} = \frac{\sqrt{\{(x'-x)^2+(y'-y)^2+(z'-z)^2\}}}{f} = 1;$$

from which it follows, that the second body will also receive a force λ directed perpendicular to the surface whose equation is $\delta u = \delta f = 0$, supposing x', y', and z' alone to vary. This surface is that of a sphere, having f for its radius, the co-ordinates x, y, and z of the first body corresponding to its centre; consequently the force that acts upon the second body, will be also directed along the thread f, which joins this body to the first.

With respect to the second body, we also have
$$\frac{\delta u'}{\delta x'} = -\frac{x''-x'}{g}, \quad \frac{\delta u'}{\delta y'} = -\frac{y''-y'}{g}, \quad \frac{\delta u'}{\delta z'} = -\frac{z''-z'}{g};$$
therefore
$$\sqrt{\left\{\left(\frac{\delta u'}{\delta x'}\right)^2 + \left(\frac{\delta u'}{\delta y'}\right)^2 + \left(\frac{\delta u'}{\delta z'}\right)^2\right\}} = 1.$$

The second body will therefore be acted upon by a force equal to λ', the direction of which will be perpendicular to

we shall have $\delta f = 0$, $\delta f' = 0$, $\delta f'' = 0$ &c.; the necessary conditions will therefore be fulfilled, and in con-

---

the surface represented by the equation $\delta u' = 0$, by making x', y', and z' to vary. This surface is spherical, having g for its radius; therefore the direction of the force λ' will be along this radius, that is, along the line which joins the second and third bodies.

Similar conclusions may be drawn with respect to the other bodies.

It is evident that the force λ which acts upon the first body, along the direction of the thread which joins it to the next, and the equal but directly contrary force λ, which acts upon the second body along the direction of the same thread, are merely the forces resulting from the re-action of this thread upon the two bodies, that is, from the tension of that portion of the thread which is included between the first and second bodies; therefore the co-efficient λ will express the force of this tension. In like manner, the co-efficient λ' will express the tension of that part of the thread which is intercepted between the second and third bodies, and so on with the rest.

It has been supposed in the solution of this problem, that each portion of the thread was not only inextensible, but likewise incompressible, so that it always preserved the same length, consequently the forces λ, λ', &c. only express the tensions when they are positive, and their actions incline the bodies towards each other; but if they are negative, and tend to make them separate to a greater distance from each other, they rather express the resistances which the thread opposes to the bodies by means of its stiffness or incompressibility.

To confirm what has been demonstrated, and at the same time to give a new application of these methods, we will sup-

sequence of the equation (*l*) the following may be obtained

---

pose that the thread to which the bodies are attached is elastic, and susceptible of extension and contraction, and that F, G, &c. are the forces of contraction of the portions f, g, &c. of the thread intercepted between the first and the second bodies, and between the second and the third &c.

It is evident from what has preceded, that the forces F, G, &c. will give the moments $F.\delta f + G.\delta g + \&c.$ or $\lambda.\delta u + \lambda'.\delta u' + \&c.$

It is therefore necessary to add these moments to those which arise from the action of the forces which are represented by the formula $P.\delta x + Q.\delta y + R.\delta z + P'.\delta x' + Q'.\delta y' + R'.\delta z' + P''.\delta x'' + \&c.$; and as there are no other particular conditions to fulfil relative to the situation of the bodies, the general equation of equilibrium will be as follows $P.\delta x + Q.\delta y + R.\delta z + P'.\delta x' + Q'.\delta y' + R'.\delta z' + P''.\delta x'' + \&c. + \lambda.\delta u + \lambda'.\delta u' + \&c. = 0$.

By substituting the values of $\delta f, \delta g$, &c. found above, and equalling to nothing the sum of the terms affected by each of the differentials $\delta x, \delta y$, &c. we shall have the following equations for the equilibrium of the thread in the present case,

$$P - \frac{F(x'-x)}{f} = 0,$$

$$Q - \frac{F(y'-y)}{f} = 0,$$

$$R - \frac{F(z'-z)}{f} = 0;$$

$$P' + \frac{F(x'-x)}{f} - \frac{G(x''-x')}{g} = 0,$$

$$Q' + \frac{F(y'-y)}{f} - \frac{G(y''-y')}{g} = 0,$$

$$0 = \Sigma . mS. \left(\frac{\delta s}{\delta x}\right); \quad 0 = \Sigma . mS. \left(\frac{\delta s}{\delta y}\right); \quad 0 = \Sigma . mS. \left(\frac{\delta s}{\delta z}\right); \quad (m)$$

$$R' + \frac{F(z'-z)}{f} - \frac{G(z''-z')}{g} = 0,$$

$$P'' + \frac{G(x''-x')}{g} = 0,$$

$$Q'' + \frac{G(y''-y')}{g} = 0,$$

$$R'' + \frac{G(z''-z')}{g} = 0.$$

These equations are analogous to those of the case in which the thread is inextensible, and give by comparison $\lambda = F$, $\lambda' = G$, &c.

It therefore appears that the quantities F, G, &c. which here express the forces of threads supposed elastic, are the same as those which have been found before to express the forces of the same threads, on the supposition that they were inextensible.

Let us resume the case of an inextensible thread charged with three bodies, supposing at the same time that the second body is moveable along the thread; in this case the condition of the problem will be, that the sum of the distances between the first and second bodies, and between the second and third is constant: denoting as before, by f and g, these

which are three of the six equations that contain the conditions of the equilibrium of the system. The se-

---

distances, we shall have $f+g$ equal to a constant quantity, and consequently $\delta f + \delta g = 0$.

In this case $\delta f + \delta g = \delta u$, and consequently $\lambda(\delta f + \delta g)$ or $\lambda \delta u$ must be added to the general equation of equilibrium, which will become

$P.\delta x + Q.\delta y + R.\delta z + P'.\delta x' + Q'.\delta y' + R'.\delta z' + P''.\delta x'' + Q''.\delta y'' + R''.\delta z'' + \lambda.\delta u = 0.$

If the values of δf and δg are substituted, and the sum of the terms which are multiplied by the differentials δx, δy, &c. equalled to nothing, the following equations, which are sufficient for the equilibrium of the thread, will be obtained,

$$P - \lambda.\frac{x'-x}{f} = 0,$$

$$Q - \lambda.\frac{y'-y}{f} = 0,$$

$$R - \lambda.\frac{z'-z}{f} = 0,$$

$$P' + \lambda\left(\frac{x'-x}{f} - \frac{x''-x'}{g}\right) = 0$$

$$Q' + \lambda\left(\frac{y'-y}{f} - \frac{y''-y'}{g}\right) = 0,$$

$$R' + \lambda\left(\frac{z'-z}{f} - \frac{z''-z'}{g}\right) = 0,$$

$$P'' + \lambda.\frac{x''-x'}{g} = 0,$$

cond members of these equations are the sums of the forces of the system resolved parallel to the three axes

$$Q'' + \lambda . \frac{y'' - y'}{g} = 0,$$

$$R'' + \lambda . \frac{z'' - z'}{g} = 0.$$

It is only necessary to extract the indeterminate quantity λ from these equations.

From the above examples it is easy to perceive how we may extend the question to a greater number of bodies, of which the end ones may be supposed fixed, and the others moveable along the thread.

Let us now suppose that the three bodies are united by inflexible rods, and obliged always to remain at the same distance from each other; in this case, if h be supposed to denote the distance between the first and third bodies, we shall have $\delta f = 0$, $\delta g = 0$, and $\delta h = 0$; consequently by having three indeterminate co-efficients, the general equation of equilibrium will become

$$P.\delta x + Q.\delta y + R.\delta z + P'.\delta x' + Q'.\delta y' + R'.\delta z' + P''.\delta x'' + Q''.\delta y'' + R''.\delta z'' + \lambda.\delta u + \lambda'.\delta u' + \lambda''.\delta u'' = 0.$$

The values of δf, and δg, or δu and $\delta u'$ have been given before; that of δh, as

$$h = \sqrt{\{(x'' - x)^2 + (y'' - y)^2 + (z'' - z)^2\}},$$

will be

$$\delta h = \frac{(x'' - x)(\delta x'' - \delta x) + (y'' - y)(\delta y'' - \delta y) + (z'' - z)(\delta z'' - \delta z)}{h}$$

By making these substitutions, and equalling to nothing the sum of the terms affected by each of the differentials δx, δy, &c. we shall obtain the nine following particular equations

x, y, and z: each of which sums should be equal to nothing in the case of equilibrium.

$$P - \lambda \cdot \frac{x'-x}{f} - \lambda'' \cdot \frac{x''-x}{h} = 0,$$

$$Q - \lambda \cdot \frac{y'-y}{f} - \lambda'' \cdot \frac{y''-y}{h} = 0,$$

$$R - \lambda \cdot \frac{z'-z}{f} - \lambda'' \cdot \frac{z''-z}{h} = 0,$$

$$P' + \lambda \cdot \frac{x'-x}{f} - \lambda' \cdot \frac{x''-x'}{g} = 0,$$

$$Q' + \lambda \cdot \frac{y'-y}{f} - \lambda' \cdot \frac{y''-y'}{g} = 0,$$

$$R' + \lambda \cdot \frac{z'-z}{f} - \lambda' \cdot \frac{z''-z'}{g} = 0,$$

$$P'' + \lambda' \cdot \frac{x''-x'}{g} + \lambda'' \cdot \frac{x''-x}{h} = 0,$$

$$Q'' + \lambda' \cdot \frac{y''-y'}{g} + \lambda'' \cdot \frac{y''-y}{h} = 0,$$

$$R'' + \lambda' \cdot \frac{z''-z'}{g} + \lambda'' \cdot \frac{z''-z}{h} = 0.$$

It will be necessary to extract from them the three unknown indeterminate quantities λ, λ', and λ'', by which means six equations will be obtained to determine the conditions of equilibrium.

It is evident from the form of these equations, that by adding respectively the three first to the three next, and afterwards to the three last, the three following equations will be obtained, which are free from the quantities λ, λ', and λ'',

$$P + P' + P'' = 0,$$
$$Q + Q' + Q'' = 0,$$
$$R + R' + R'' = 0.$$

It would be very easy to find three other equations by the

The equations $\delta f=0$, $\delta f'=0$, $\delta f''=0$, &c. will be also satisfied, if we suppose z, z', z'', &c. invariable quantities, and if we make

---

extraction of λ, λ', and λ''; but this may be done in a much readier and more general manner, by deducing these nine equations from those given above.

$$Py - Qx - \lambda \cdot \frac{yx' - xy'}{f} - \lambda'' \cdot \frac{yx'' - xy''}{h} = 0,$$

$$Pz - Rx - \lambda \cdot \frac{zx' - xz'}{f} - \lambda'' \cdot \frac{zx'' - xz''}{h} = 0,$$

$$Qz - Ry - \lambda \cdot \frac{zy' - yz'}{f} - \lambda'' \cdot \frac{zy'' - yz''}{h} = 0,$$

$$P'y - Q'x + \lambda \cdot \frac{yx' - xy'}{f} - \lambda' \cdot \frac{y'x'' - x'y''}{g} = 0,$$

$$P'z - R'x + \lambda \cdot \frac{zx' - xz'}{f} - \lambda' \cdot \frac{z'x'' - x'z''}{g} = 0,$$

$$Q'z - R'y + \lambda \cdot \frac{zy' - yz'}{f} - \lambda' \cdot \frac{z'y'' - y'z''}{g} = 0,$$

$$P''y'' - Q''x'' + \lambda' \cdot \frac{y'x'' - x'y''}{g} + \lambda'' \cdot \frac{yx'' - xy''}{h} = 0,$$

$$P''z'' - R''x'' + \lambda' \cdot \frac{z'x'' - x'z''}{g} + \lambda'' \cdot \frac{zx'' - xz''}{h} = 0,$$

$$Q''z'' - R''y'' + \lambda' \cdot \frac{z'y'' - y'z''}{g} + \lambda'' \cdot \frac{zy'' - yz''}{h} = 0.$$

These equations are evidently analogous to the primitive ones, and give in the same manner, by addition, the three following equations,

$$Py - Qx + P'y' - Q'x' + P''y'' - Q''x'' = 0,$$
$$Pz - Rx + P'z' - R'x' + P''z'' - R''x'' = 0,$$
$$Qz - Ry + Q'z' - R'y' + Q''z'' - R''y'' = 0.$$

$$\delta x = y.\delta\varpi\ ; \qquad \delta y = -x.\delta\varpi\ ;$$
$$\delta x' = y'.\delta\varpi\ ; \qquad \delta y' = -x'.\delta\varpi\ ;$$
$$\&c. \qquad\qquad \&c.$$

The three first equations shew, that the sum of the forces parallel to each of the three axes should be nothing, and the three last contain the known principle of moments (understanding by that term the product of the power by its distance, as in the case of the lever), from which it appears, that the sum of the moments of all the forces to make the system turn about each of the three axes, is likewise nothing.

If the first body be fixed, the differentials δx, δy, and δz will vanish, and the three first of the nine equations first given will not have place; we shall in this case have only six equations, which by the extraction of the three indeterminates λ, λ', and λ'', may be reduced to three.

In order to obtain these three equations we may proceed in a manner analogous to that made use of to discover the three last equations of the preceding question; provided we take care that the transformed ones do not contain the indeterminates λ, and λ'', which enter into the three first, of which it is necessary to make abstraction. This will be obtained by the following combinations,

$$P'(y'-y)-Q'(x'-x)-\lambda'\frac{(y'-y)(x''-x')-(x'-x)(y''-y)}{g}=0,$$

$$P'(z'-z)-R'(x'-x)-\lambda'.\frac{(z'-z)(x''-x')-(x'-x)(z''-z')}{g}=0,$$

$$Q'(z'-z)-R'(y'-y)-\lambda'.\frac{(z'-z)(y''-y')-(y'-y)(z''-z')}{g}=0,$$

$$P''(y''-y)-Q''(x''-x)+\lambda'\frac{(y''-y)(x''-x')-(x''-x)(y''-y')}{g}=0,$$

$$P''(z''-z)-R''(x''-x)+\lambda'.\frac{(z''-z)(x''-x')-(x''-x)(z''-z')}{g}=0,$$

$\delta\varpi$ being any variation whatever. By substituting these values in the equation *(l)*, we shall have

$$0 = \Sigma . m S . \left\{ y . \left(\frac{\delta s}{\delta x}\right) - x . \left(\frac{\delta s}{\delta y}\right) \right\}.$$

$$Q''(z''-z) - R''(y''-y) + \lambda' \frac{(z''-z)(y''-y') - (y''-y)(z''-z')}{s} = 0.$$

If we add the three first of these equations to the three last, the three following will be obtained,

$$P'(y'-y) - Q'(x'-x) + P''(y''-y) - Q''(x''-x) = 0,$$
$$P'(z'-z) - R'(x'-x) + P''(z''-z) - R''(x''-x) = 0,$$
$$Q'(z'-z) - R'(y'-y) + Q''(z''-z) - R''(y''-y) = 0.$$

These will always have place, whatever may be the state of the first body, as they are independent of the equations relative to it. These equations contain the principle of moments, with respect to the axes passing through the first body.

Let us suppose a fourth body attached to the same inflexible rod, having x''', y''', and z''' for its three rectangular co-ordinates, and P''', Q''', and R''' for the three forces parallel to these co-ordinates.

It will in this case, be necessary to add the quantity $P'''.\delta x''' + Q'''.\delta y''' + R'''.\delta z'''$ to the sum of the moments of the forces. As the distances between all the bodies ought to remain constant, we shall have by the conditions of the problem, not only $\delta f = 0$, $\delta g = 0$, $\delta h = 0$, as in the preceding case, but also $\delta l = 0$, $\delta m = 0$, and $\delta n = 0$; naming the distances of the fourth body from the three others l, m, and n. The general equation of equilibrium will in this case become

$$P.\delta x + Q.\delta y + R.\delta z + P'.\delta x' + Q'.\delta y' + R'.\delta z' + P''.\delta x'' + Q''.\delta y'' + R''.\delta z'' + P'''.\delta x''' + Q'''.\delta y''' + R'''.\delta z''' + \lambda.\delta u + \lambda'.\delta u' + \lambda''.\delta u'' + \lambda'''.\delta u''' + \lambda^{iv}.\delta u^{iv} + \lambda^{v}.\delta u^{v} = 0.$$

It is evident that we may in this equation change either the co-ordinates x, x', x'', &c. or y, y', y'', &c. into

---

The values of δf, δg, and δh, are the same as before, and those of δl, δm, and δn, or $\delta u'''$, δu^{iv}, and δu^{v}, as

$$l = \sqrt{\{(x'''-x)^2+(y'''-y)^2+(z'''-z)^2\}},$$
$$m = \sqrt{\{(x'''-x')^2+(y'''-y')^2+(z'''-z')^2\}},$$
$$n = \sqrt{\{(x'''-x'')^2+(y'''-y'')^2+(z'''-z'')^2\}},$$

are

$$\delta u''' = \delta l = \frac{(x'''-x)(\delta x'''-\delta x)+(y'''-y)(\delta y'''-\delta y)+(z'''-z)(\delta z'''-\delta z)}{l}$$

$$\delta u^{iv} = \delta m = \frac{(x'''-x')(\delta x'''-\delta x')+(y'''-y')(\delta y'''-\delta y')+(z'''-z')(\delta z'''-\delta z')}{m}$$

$$\delta u^{v} = \delta n = \frac{(x'''-x'')(\delta x'''-\delta x'')+(y'''-y'')(\delta y'''-\delta y'')+(z'''-z'')(\delta z'''-\delta z'')}{n}$$

By making these substitutions, and equalling to nothing the sum of the terms of each of the differentials δx, δy, &c., we shall find twelve particular equations; the nine first of which which will be the same as those in the case of three bodies, if the following quantities were respectively added to their first members.

$$-\lambda'''.\frac{x'''-x}{l}, \quad -\lambda'''.\frac{y'''-y}{l}, \quad -\lambda'''.\frac{z'''-z}{l},$$

$$-\lambda_{iv}.\frac{x'''-x'}{m}, \quad -\lambda_{iv}.\frac{y'''-y'}{m}, \quad -\lambda_{iv}.\frac{z'''-z'}{m},$$

$$-\lambda^{v}.\frac{x'''-x''}{n}, \quad -\lambda^{v}.\frac{y'''-y''}{n}, \quad -\lambda^{v}.\frac{z'''-z''}{n};$$

and the three last will be

$$P''' + \lambda'''.\frac{x'''-x}{l} + \lambda^{iv}.\frac{x'''-x'}{m} + \lambda^{v}.\frac{x'''-x''}{n} = 0,$$

$$Q''' + \lambda'''.\frac{y'''-y}{l} + \lambda^{iv}.\frac{y'''-y'}{m} + \lambda^{v}.\frac{y'''-y''}{n} = 0,$$

z, z', z'', &c. which will give two other equations that re-united to the preceding, will form the following system

$$0 = \Sigma.mS.\left\{ y.\left(\frac{\delta s}{\delta x}\right) - x.\left(\frac{\delta s}{\delta y}\right) \right\}$$
$$0 = \Sigma.mS.\left\{ z.\left(\frac{\delta s}{\delta x}\right) - x.\left(\frac{\delta s}{\delta z}\right) \right\} \quad ; (n)$$
$$0 = \Sigma.mS.\left\{ y.\left(\frac{\delta s}{\delta z}\right) - z.\left(\frac{\delta s}{\delta y}\right) \right\}$$

$$R''' + \lambda'''.\frac{z''' - z}{l} + \lambda^{iv}.\frac{z''' - z'}{m} + \lambda^{v}.\frac{z''' - z''}{n} = 0.$$

As there are twelve equations in all, and six indeterminate quantities λ, λ', λ'', λ''', λ^{iv}, λ^{v}, to eliminate, there will only remain six final equations for the conditions of equilibrium, as in the case of three bodies; and we shall find by a method similar to one given before, these six equations analogous to those found in that case,

$$P + P' + P'' + P''' = 0,$$
$$Q + Q' + Q'' + Q''' = 0,$$
$$R + R' + R'' + R''' = 0,$$
$$Py - Qx + P'y' - Q'x' + P''y'' - Q''x'' + P'''y''' - Q'''x'''$$
$$= 0,$$
$$Pz - Rx + P'z' - R'x' + P''z'' - R''x'' + P'''z''' - R'''x'''$$
$$= 0,$$
$$Qz - Ry + Q'z' - R'y' + Q''z'' - R''y'' + Q'''z''' - R'''y'''$$
$$= 0.$$

Instead of the three last, the three following equations may be substituted, which can be found by a method given before. As they are independent of the equations relative to the first body, they possess the advantage of always having place, whatever may be the state of this body.

the function $\Sigma.mS.y.\left(\frac{\delta s}{\delta x}\right)$] is by No. 3, the sum of the moments of all the forces parallel to the axis of x,

---

$$P'(y'-y)-Q'(x'-x)+P''(y''-y)-Q''(x''-x)+$$
$$P'''(y'''-y)-Q'''(x'''-x)=0,$$

$$P'(z'-z)-R'(x'-x)+P''(z''-z)-R''(x''-x)+$$
$$P'''(z'''-z)-R'''(x'''-x)=0,$$

$$Q'(z'-z)-R'(y'-y)+Q''(z''-z)-R''(y''-y)+$$
$$Q'''(z'''-z)-R'''(y'''-y)=0.$$

Let us now consider the case of three bodies joined by a rod which is elastic at the point where the second body is situated, the distances between it and the other bodies being constant, but the angles which the lines form variable. Let us suppose that the force of elasticity, which tends to augment the angle formed by the lines which join the second body to the two others, is represented by E, and the exterior angle, formed by one of these sides and the prolongation of the other, by e; then the moment of the force E ought to be represented by $E.\delta e$, or its equal $\lambda''.\delta u''$; therefore the sum of the moments of all the forces of the system, as $\delta f=0$, $\delta g=0$, will be

$$P.\delta x+Q.\delta y+R.\delta z+P'.\delta x'+Q'.\delta y'+R'.\delta z'+P''.\delta x''$$
$$Q''.\delta y''+R''.\delta z''+\lambda.\delta u+\lambda'.\delta u'+\lambda''.\delta u''=0.$$

It is now only required to substitute the values of δu, $\delta u'$, and $\delta u''$: those of δu and $\delta u'$ are the same as in the first question, but with respect to that of $\delta u''$ or δe, it may be observed, that in the triangle, of which the three sides are f, g, and h, or the distance of the first body from the third, $180-e$ is the angle opposite to the side h; therefore by

which would cause the system to revolve about the axis of z. In like manner, the function $\Sigma . m S . x . \left(\frac{\delta s}{\delta y}\right)$ is

---

trigonometry $-\cos. e = \frac{f^2 + g^2 - h^2}{2fg}$; which by differentiation gives the value of δe or $\delta u''$: as by the conditions of the problem $\delta f = 0$, $\delta g = 0$, it will be sufficient to make e and h vary, we shall therefore have $\delta e = \delta u'' = -\frac{h \delta h}{fg \sin . e}$. This value being substituted in the preceding equation, it will evidently become of the same form as the general equation of equilibrium given in the case of three bodies joined together by an inflexible rod; by supposing in it that $\lambda'' = -\frac{Eh}{fg \sin . e}$; the particular equations will necessarily be the same in the two cases, with this sole difference, that in the case above mentioned, the quantity λ'' is indeterminate, and consequently ought to be eliminated; but in the present case it is known, and there are only two quantities λ and λ' to eliminate, consequently there will be seven final equations instead of six. But whether the quantity λ'' is known or not, it may be eliminated along with the two others, λ and λ'; we shall therefore have, in the present case, the same equations as were found in the case of three bodies attached to each other by an inflexible rod: to find the seventh equation it will only be necessary to eliminate λ from the three first, or λ' from the three last of the nine particular equations of the above case, and to substitute for λ'' it value $-\frac{Eh}{fg \sin . e}$.

If δf and δg had not been supposed to vanish in the value

the sum of the moments of all the forces parallel to the axis of y, which would cause the system to turn about the axis of z, but in a different direction to the first forces; the first of the equations (n) consequently indicates, that the sum of the moments of the forces,is

---

of δe, it would have been of this form $\delta e = -\dfrac{h \delta h}{fg \sin. e} + A.\delta f + B.\delta g$, A and B being functions of f, g, h, and $\sin. e$; in this case, the three terms $\lambda.\delta u + \lambda'.\delta u' + \lambda''.\delta u''$ of the general equation, would become $(EA+\lambda).\delta u + (EB+\lambda').\delta u' - \dfrac{Eh}{fg \sin. e}.\delta h$; but λ and λ' being two indeterminate quantities, it is evident that $\lambda - EA$ and $\lambda' - EB$ may be substituted for them, by which means the quantity treated upon will become $\lambda.\delta u + \lambda'.\delta u' - \dfrac{Eh}{fg \sin. e}.\delta h$, the same as when f and g did not vary in the expression of δe.

If many bodies be supposed to be joined together by elastic rods, we shall find, in the same manner, the proper equations for the equilibrium of these bodies. The above methods will always give with the same facility, the conditions of the equilibrium of a system of bodies connected together in any manner, and acted upon by any exterior forces whatever. The proceedings are always similar, which ought to be regarded as one of the principal advantages of this method.

The following question, and several others, are likewise solved in the Mechanique Analytique of Lagrange. To find the equilibrium of a thread, all the points of which are acted upon by any forces whatever, and which is supposed perfectly flexible or inflexible, extensible or inextensible, elastic or inelastic.

nothing with respect to the axis of z*. The second and the third of these equations indicate in a similar man-

---

* If the system be at liberty to turn in any direction about a point taken for the origin of the co-ordinates, the instantaneous rotations about the three axes, may be considered in the following manner; which will give three equations of rotation with respect to these axes, similar to those of Laplace.

Let ρ represent the projection of a line drawn from the origin of the co-ordinates to the body m, ρ' that drawn with respect to the body m', &c., also φ the angle which the line ρ makes with the axis of x, φ' that which ρ' makes with the same axis, &c. we shall have the following equations $x = \rho . \cos . \varphi$, $y = \rho . \sin . \varphi$, $x' = \rho' . \cos . \varphi'$, $y' = \rho' . \sin . \varphi'$, &c. which by differentiation and substitution, if ρ, ρ', ρ'' &c. be supposed constant, will give

$$\delta x = -y . \delta \varphi, \quad \delta y = x . \delta \varphi, \quad \delta x' = -y' . \delta \varphi, \quad \delta y' = x' . \delta \varphi, \text{ &c.}$$

$\delta \varphi'$, $\delta \varphi''$, &c. being each of them supposed equal to $\delta \varphi$, as the bodies m, m', m'', &c. are imagined to be invariably connected.

These variations of x, y, x', y', &c. are owing to the elementary rotation $\delta \varphi$ about the axis of z.

In a similar manner, if ψ, ψ', ψ'', &c. represent the angles which the projections upon the plane yz of lines drawn from the centre of the co-ordinates to the bodies m, m', m'', &c. respectively make with the axis y, the variations of y, z, y', z', &c. arising from the elementary rotation $\delta \psi$ about the axis of x may be obtained, which will give the following equations,

$$\delta y = -z . \delta \psi, \quad \delta z = y . \delta \psi, \quad \delta y' = -z' . \delta \psi, \quad \delta z' = y' . \delta \psi, \text{ &c.}$$

Likewise if ω, ω', ω'', &c. represent the angles which the projections upon the plane xz of lines drawn from the centre of the co-ordinates to the bodies m, m', m'', &c. make

&c, that the sum of the moments of the forces is nothing, either with respect to the axis of y, or to the axis

with the axis z, the variations of x, z, x', z', &c., arising from the elementary rotation $\delta\omega$ about the axis of y, will give
$$\delta z = -x.\delta\omega,\ \delta x = z.\delta\omega,\ \delta z' = -x'.\delta\omega,\ \delta x' = z'.\delta\omega,\ \&c.$$
If the three rotations take place at the same time, the whole variations of the co-ordinates x, y, z, x', y', z', &c. will be equal to the sums of the partial variations belonging to each of these rotations, consequently we shall have the following equations,
$$\delta x = z.\delta\omega - y.\delta\varphi,\ \delta y = x.\delta\varphi - z.\delta\psi,\ \delta z = y.\delta\psi - x.\delta\omega,$$
$$\delta x' = z'.\delta\omega - y'.\delta\varphi,\ \delta y' = x'.\delta\varphi - z'.\delta\psi,\ \delta z' = y'.\delta\psi - x'.\delta\omega.$$
If these values be substituted in the general formula of equilibrium $\Sigma.S.\delta s = 0$, we shall obtain the terms belonging to the rotations $\delta\varphi$, $\delta\omega$, and $\delta\psi$ about the the three axes of z, y, and x; which ought to be separately equal to nothing, when the system has liberty to turn in any direction about a point placed at the origin of the co-ordinates. The equation $\Sigma.S.\delta s = 0$, by substitution, gives the following
$$L.\delta\psi + M.\delta\omega + N.\delta\varphi = 0.$$
in which
$$L = \Sigma.mS.\left\{y.\left(\frac{\delta s}{\delta z}\right) - z.\left(\frac{\delta s}{\delta y}\right)\right\},$$
$$M = \Sigma.mS.\left\{z.\left(\frac{\delta s}{\delta x}\right) - x.\left(\frac{\delta s}{\delta z}\right)\right\},$$
$$N = \Sigma.mS.\left\{x.\left(\frac{\delta s}{\delta y}\right) - y.\left(\frac{\delta s}{\delta x}\right)\right\}.$$
The co-efficients of the instantaneous rotations $\delta\psi$, $\delta\omega$, and $\delta\varphi$, are the moments relative to the axes of these rotations, and are respectively equal to nothing in the case of equilibrium, when the system has liberty to turn about the origin of the co-ordinates.

of x. By uniting these three conditions to those in

If at any point of the system the co-ordinates x, y, and z be respectively proportional to $\delta\psi$, $\delta\omega$, and $\delta\varphi$, we shall have there
$$z.\delta\omega = y.\delta\varphi, \quad x.\delta\varphi = z.\delta\psi, \quad y.\delta\psi = x.\delta\omega,$$
and consequently $\delta x = 0$, $\delta y = 0$, and $\delta z = 0$.

This point and all others which have the same property, will consequently be immoveable during the instant that the system describes the three angles $\delta\psi$, $\delta\omega$, and $\delta\varphi$, by turning at the same time about the three axes of x, y and z. It may be easily proved that all the points which have this property are in a right line passing through the origin of the co-ordinates. The co-sines of the angles λ, μ, and ν which it makes with the axes of x, y and z, are
$$\frac{x}{\sqrt{x^2+y^2+z^2}}, \quad \frac{y}{\sqrt{x^2+y^2+z^2}}, \quad \text{and} \quad \frac{z}{\sqrt{x^2+y^2+z^2}},$$
that by substitution will respectively become
$$\frac{\delta\psi}{\sqrt{(\delta\psi^2+\delta\omega^2+\delta\varphi^2)}}, \quad \frac{\delta\omega}{\sqrt{(\delta\psi^2+\delta\omega^2+\delta\varphi^2)}} \quad \text{and}$$
$$\frac{\delta\varphi}{\sqrt{(\delta\psi^2+\delta\omega^2+\delta\varphi^2)}}.$$

This right line is the instantaneous axis of the composed rotation.

If we suppose $\delta\theta = \sqrt{(\delta\psi^2+\delta\omega^2+\delta\varphi^2)}$ we shall have
$$\delta\psi = \delta\theta.\cos.\lambda, \quad \delta\omega = \delta\theta.\cos.\mu, \quad \delta\varphi = \delta\theta.\cos.\nu,$$
which, by substitution in the general expressions of δx, δy, and δz, will give
$$\delta x = (z.\cos.\mu - y.\cos.\nu)\delta\theta,$$
$$\delta y = (x.\cos.\nu - z.\cos.\lambda)\delta\theta,$$
$$\delta z = (y.\cos.\lambda - x.\cos.\mu)\delta\theta.$$

These values being substituted in the expression $\delta x^2 + \delta y^2 + \delta z^2$, which is the square of the indefinitely short space passed

which the sum of the forces parallel to these axis are

---

ever by any point whatever, it will be changed into the following

$$\{(z\cos.\mu - y\cos.\nu)^2 + (x\cos.\nu - z\cos.\lambda)^2 + (y\cos.\lambda - x\cos.\mu)^2\}$$
$$.\delta\theta^2 = \{x^2 + y^2 + z^2 - (x\cos.\lambda + y\cos.\mu + z\cos.\nu)^2\}\delta\theta^2$$

as $\cos.^2\lambda + \cos.^2\mu + \cos.^2\nu = 1$.

It may readily be proved that $x\cos.\lambda + y\cos.\mu + z\cos.\nu = 0$, is the equation to a plane passsing through the origin of the co-ordinates, in a direction perpendicular to the right line which makes the angles λ, μ, and ν, with the axes of x, y, and z, consequently the short space described by any point of this plane will be $\delta\theta\sqrt{x^2 + y^2 + z}$. As the axis of rotation is perpendicular to this plane, $\delta\theta$ will represent the angle of rotation about it composed of the three partial velocities $\delta\psi, \delta\omega$, and $\delta\varphi$, about the three axes of the co-ordinates.

It therefore follows, that any instantaneous rotations $\delta\psi$, $\delta\omega$, and $\delta\varphi$ about three axes which cut each other at right angles at the same point, may be composed into one $\delta\theta = \sqrt{\delta\psi^2 + \delta\omega^2 + \delta\varphi^2}$ about an axis passing through the same point of intersection, and making with them the angles λ, μ, and ν. so that

$$\cos.\lambda = \frac{\delta\psi}{\delta\theta}, \quad \cos.\mu = \frac{\delta\omega}{\delta\theta}, \quad \cos.\nu = \frac{\delta\varphi}{\delta\theta}.$$

Inversely, any rotation $\delta\theta$ about a given axis may be resolved into three partial rotations, denoted by $\cos.\lambda.\delta\theta$, $\cos.\mu.\delta\theta$, and $\cos.\nu \delta\theta$, about three axes, which cut each other perpendicularly in a point of the given axis, and make with it the angles λ, μ, and ν. The above enable us in a very easy manner to compose and to resolve the instantaneous movements or velocities of rotation.

Let three other rectangular axes be taken which make with the axis of rotation $\delta\psi$ the angles λ', λ''. and λ''', with the axis of rotation $\delta\omega$ the angles μ', μ'', μ'''. and with the axis of rota-

nothing with respect to each of them; we shall have

---

tion $\delta\varphi$ the angles ν', ν'', and ν'''; the rotation $\delta\psi$ may be resolved into three rotations $\cos.\lambda'.\delta\psi$, $\cos.\lambda''.\delta\psi$, and $\cos.\lambda'''.\delta\psi$ about these new axes, the rotation $\delta\omega$ may likewise be resolved into three rotations $\cos.\mu'.\delta\omega$, $\cos.\mu''.\delta\omega$, and $\cos.\mu'''.\delta\omega$, and the rotation $\delta\varphi$ into three rotations $\cos.\nu'.\delta\varphi$, $\cos.\nu''.\delta\varphi$, and $\cos.\nu'''.\delta\varphi$ about the same axes. By adding together the rotations about the same axis, if we name $\delta\theta'$, $\delta\theta''$, and $\delta\theta'''$ the complete rotations about the three new axes we shall have

$$\delta\theta' = \cos.\lambda'.\delta\psi + \cos.\mu'.\delta\omega + \cos.\nu'.\delta\varphi,$$
$$\delta\theta'' = \cos.\lambda''.\delta\psi + \cos.\mu''.\delta\omega + \cos.\nu''.\delta\varphi,$$
$$\delta\theta''' = \cos.\lambda'''.\delta\psi + \cos.\mu'''.\delta\omega + \cos.\nu'''.\delta\varphi.$$

The rotations $\delta\psi$, $\delta\omega$, and $\delta\varphi$, are by this means reduced to three rotations $\delta\theta'$, $\delta\theta''$, and $\delta\theta'''$ about three other rectangular axes, which should consequently give the same rotation $\delta\theta$ that results from the rotations $\delta\psi$, $\delta\omega$, and $\delta\varphi$, we shall therefore have

$$\delta\theta^2 = \delta\theta'^2 + \delta\theta''^2 + \delta\theta'''^2 = \delta\psi^2 + \delta\omega^2 + \delta\varphi^2;$$

as this equation is identic, by substituting for $\delta\theta'^2$, $\delta\theta''^2$, $\delta\theta'''^2$ their values given above, the following conditional equations will be obtained,

$$\cos.^2\lambda' + \cos.^2\lambda'' + \cos.^2\lambda''' = 1,$$
$$\cos.^2\mu' + \cos.^2\mu'' + \cos.^2\mu''' = 1,$$
$$\cos.^2\nu' + \cos.^2\nu'' + \cos.^2\nu''' = 1;$$
$$\cos.\lambda'.\cos.\mu' + \cos.\lambda''\cos.\mu'' + \cos.\lambda'''.\cos.\mu''' = 0,$$
$$\cos.\lambda'.\cos.\nu' + \cos.\lambda''.\cos.\nu'' + \cos.\lambda'''.\cos.\nu''' = 0,$$
$$\cos.\mu'.\cos.\nu' + \cos.\mu''.\cos.\nu'' + \cos.\mu'''.\cos.\nu''' = 0.$$

The three first are the respective co-efficients of $\delta\psi$, $\delta\omega$, and $\delta\varphi$, which must each of them be equal to unity, and the three last the respective co-efficients of $2\delta\theta'.\delta\psi$, $2\delta\theta''.\delta\omega$, and $2\delta\theta'''.\delta\varphi$, and consequently should each of them vanish that the equation may be identic.

the six conditions of the equilibrium of a system of bodies invariably connected together.

By means of these relations, the values of $\delta\psi$, $\delta\omega$, and $\delta\varphi$ may be obtained, in terms of $\delta\theta'$, $\delta\theta''$, and $\delta\theta'''$, by adding together the values of $\delta\theta'$, $\delta\theta''$, and $\delta\theta'''$, multiplied successively by $\cos.\lambda'$, $\cos.\lambda''$, $\cos.\lambda'''$, $\cos.\mu'$, $\cos.\mu''$, &c., which will give

$$\delta\psi = \cos.\lambda'.\delta\theta' + \cos.\lambda''\delta\theta'' + \cos.\lambda'''\delta\theta''',$$
$$\delta\omega = \cos.\mu'.\delta\theta' + \cos.\mu''.\delta\theta'' + \cos.\mu'''.\delta\theta''',$$
$$\delta\varphi = \cos.\nu'.\delta\theta' + \cos.\nu''.\delta\theta'' + \cos.\nu'''.\delta\theta'''.$$

If the angles which the composed rotation $\delta\theta$ makes with the axes of the three partial rotations $\delta\theta'$, $\delta\theta''$, and $\delta\theta'''$ are denoted by π, π', and π'', we shall have

$$\delta\theta' = \cos.\pi'.\delta\theta, \quad \delta\theta'' = \cos.\pi''.\delta\theta, \text{ and } \delta\theta''' = \cos.\pi'''.\delta\theta,$$

and if in the before given values of $\delta\theta'$, $\delta\theta''$, and $\delta\theta'''$, there are substituted for $\delta\psi$, $\delta\omega$, and $\delta\varphi$, their values $\cos.\lambda.\delta\theta$, $\cos.\mu.\delta\theta$, and $\cos.\nu.\delta\theta$; the comparison of these different expressions of $\delta\theta'$, $\delta\theta''$, and $\delta\theta'''$ will give, when divided by $\delta\theta$, the following new conditional equations, which may be geometrically demonstrated. Vide No. 29 Notes.

$$\cos.\pi' = \cos.\lambda.\cos.\lambda' + \cos.\mu.\cos.\mu' + \cos.\nu.\cos.\nu',$$
$$\cos.\pi'' = \cos.\lambda.\cos.\lambda'' + \cos.\mu.\cos.\mu'' + \cos.\nu.\cos.\nu'',$$
$$\cos.\pi''' = \cos.\lambda.\cos.\lambda''' + \cos.\mu.\cos.\mu''' + \cos.\nu.\cos.\nu'''.$$

The above proof shews that the compositions and the resolutions of the movements of rotation are analogous to those of rectilinear motions. For, if upon the three axes of the rotations of $\delta\psi$, $\delta\omega$, and $\delta\varphi$, from their point of intersection, three lines be taken respectively proportional to $\delta\psi$, $\delta\omega$, and $\delta\varphi$, and a rectangular parallelepiped be constructed upon them, it is evident that the diagonal of this parallelepiped will be the axis of the composed rotation $\delta\theta$, and will be at the same time proportional to this rotation.

From this, and from the consideration that the rotations about the same axis may be added to or subtracted from each

If the origin of the co-ordinates is fixed and attached

---

other, according as they are made in the same or in opposite directions, it may be concluded in general, that the composition and the resolution of the movements of rotation follow similar laws to those of the composition and the resolution of rectilinear motions, by substituting for the movements of rotation, rectilinear motions along the directions of the axes of rotation. Vide the Mechanique Analytique of Lagrange, from which the greater part of the above has been extracted.

Let $\delta\theta'$, $\delta\theta''$, $\delta\theta'''$, &c. represent any indefinite number of rotations about their respective axes, these may be composed into one $\delta\theta$, about an instantaneous axis of rotation; for if from the point where all the axes cut each other, three rectangular axes be taken, each of the rotations may be resolved into three about these axes, and by adding or subtracting, as the rotations are in the same or in contrary directions, there will be three resulting rotations which may be composed into one about an instantaneous axis of rotation. Thus if $\delta\psi'$, $\delta\omega'$, and $\delta\varphi'$ represent the three partial rotations about three rectangular co-ordinates, into which the rotation $\delta\theta'$ has been resolved, $\delta\psi''$, $\delta\omega''$, and $\delta\varphi''$, those about the same axis into which the rotation $\delta\theta''$ has been resolved, &c. the following equation may be obtained,

$$\delta\theta = \sqrt{\{(\delta\psi' + \delta\psi'' + \delta\psi''' + \&c.)^2 + (\delta\omega' + \delta\omega'' + \delta\omega''' + \&c.)^2 + (\delta\varphi' + \delta\varphi'' + \delta\varphi''' + \&c.)^2\}}.$$

If in the formula $L.\delta\psi + M.\delta\omega + N.\delta\varphi$, which contains the terms due to the rotations $\delta\psi$, $\delta\omega$, and $\delta\varphi$ in the general formula $S.\delta s + S'.\delta s' + S''.\delta s'' + \&c.$, the values of $\delta\psi$, $\delta\omega$, and $\delta\varphi$ found above be substituted, it will be changed into the following,

$$(L.\cos.\lambda' + M\cos\mu' + N.\cos.\nu')\delta\theta'$$
$$+ (L.\cos.\lambda'' + M.\cos.\mu'' + N.\cos.\nu'')\delta\theta''$$
$$+ (L.\cos.\lambda''' + M.\cos.\mu''' + N.\cos.\nu''')\delta\theta'''.$$

The co-efficients of the elementary angles $\delta\theta'$, $\delta\theta''$, and $\delta\theta'''$

invariably to the system, it will destroy the forces pa-

---

express the sums of the moments relative to the axes of the rotations $\delta\theta'$, $\delta\theta''$, and $\delta\theta'''$. From the above it appears that moments equal to L, M, and N, relative to three rectangular axes will give the moments
$$L.\cos.\lambda' + M.\cos.\mu' + N.\cos.\nu',$$
$$L.\cos.\lambda'' + M.\cos.\mu'' + N.\cos.\nu'',$$
$$L.\cos.\lambda''' + M.\cos.\mu''' + N.\cos.\nu''',$$
relative to three other rectangular axes, which respectively make with these the angles λ', μ', ν' ; λ'', μ'', ν'' ; λ''', μ''', ν'''.

A geometrical demonstration of this theorem is given by Euler in the seventh vol. of the Nova acta of the Academy of Petersburg.

If the rotations $\delta\psi$, $\delta\omega$, and $\delta\varphi$ are supposed to be proportional to L, M, and N, and we make
$$H = \sqrt{(L^2 + M^2 + N^2)},$$
the following equations will have place
$$L = H.\cos.\lambda, \quad M = H.\cos.\mu, \quad N = H.\cos.\nu,$$
and the three moments will be reduced to this simple form
$$H.\cos.\pi', \quad H.\cos.\pi'', \quad H.\cos.\pi'''.$$

But π', π'', and π''' are the angles which the axes of the rotations $\delta\theta'$, $\delta\theta''$, and $\delta\theta'''$ form with the axis of the composed rotation $\delta\theta$, if therefore we make the axis of rotation $\delta\theta'$ coincide with the axis of rotation $\delta\theta$, then $\pi' = 0$, and π'' and π''' are each equal to a right angle, consequently the moment about this axis will be H, and those about the two other axes perpendicular to it will be nothing. We may therefore conclude that moments respectively equal to L, M, and N, and relative to three rectangular axes x, y, and z may be composed into one, H, equal to $\sqrt{(L^2+M^2+N^2)}$, relative to an axis which makes with them the angles λ, μ, and ν, so that
$$\cos.\lambda = \frac{L}{H}, \quad \cos.\mu = \frac{M}{H}, \quad \cos.\nu = \frac{N}{H}.$$
The sum of the moments relative to this axis is a maximum;

rallel to the three axes; and the conditions of the equi-

---

the tangent of the angle that it makes with the plane xy is $\dfrac{N}{\sqrt{L^2 + M^2}}$, and the tangent of the angle which the projection of the axis upon that plane makes with the axis of x, is equal to $\dfrac{M}{L}$.

It is evident from the above that the composition of moments follows the same laws as that of rectilinear motions. It may be immediately deduced from the composition of instantaneous rotations, by substituting the moments for the rotations which they produce, in the same manner as forces can be substituted for right lined motions. Vide the Mechanique Analytique of Lagrange.

Those who are desirous of further information respecting the composition and the resolution of moments, may consult the writings of Euler, Prony, Poisson, &c. also a memoir by Poinsot, in the 13th Cahier of the Journal de l'Ecole Polytechnique.

If the moments of the forces which act upon a system be taken directly with respect to a point at the origin of the co-ordinates, they will follow the same laws with respect to different planes, as the projections of areas upon them, thus for instance, $Sy \cdot \left(\dfrac{\delta s}{\delta x}\right) - Sx \cdot \left(\dfrac{\delta s}{\delta y}\right)$ may No. 3 be composed into a single moment with respect to the origin of the co-ordinates. This moment will evidently be the product of the projection of the force S upon that plane multiplied by the perpendicular drawn from the origin of the co-ordinates to its direction, and may therefore be represented by an area equal to twice the area of a triangle, having the projection of a line representing in quantity and direction the force S for its base, and the origin of the co-ordinates for its summit.

It therefore follows, that the properties of moments with respect to a fixed point are similar to those of plane surfaces.

librium of a system about this origin, will be reduced

I shall mention a few circumstances concerning them, referring the reader to No. 21, and the notes accompanying it, which may be read independent of the other parts of the work, from which the following properties may be deduced.

Suppose a number of areas represented by A, A', A'', &c. are in a plane passing through the origin of the co-ordinates, let b, b', b'', &c. represent these areas projected upon three rectangular planes passing through the origin of the co-ordinates, and c, c', and c'' represent the projections of the areas upon three other rectangular planes passing through the same point, then by No. 21

$$b^2 + b'^2 + b''^2 = c^2 + c'^2 + c''^2,$$

consequently

$$b = \sqrt{(c^2 + c'^2 + c''^2 - b'^2 - b''^2)}.$$

When b' and b'' vanish, the value of b is evidently a maximum, and the line which is perpendicular to it at the origin of the co-ordinates may be found from the following equations, in which α, β, and γ represent the respective angles that it makes with the rectangular co-ordinates x, y, and z of the planes containing the areas c, c', and c''.

$$\cos.\alpha = \frac{c''}{\sqrt{(c^2 + c'^2 + c''^2)}},$$

$$\cos.\beta = \frac{c'}{\sqrt{(c^2 + c'^2 + c''^2)}},$$

$$\cos.\gamma = \frac{c}{\sqrt{(c^2 + c'^2 + c''^2)}}.$$

The absolute position of the plane of the greatest sum of the projections of the areas is indeterminate in space, as the projections are the same upon all the planes that are parallel to each other.

The sum of the projections of the areas are the same for every plane which is equally inclined to that of the greatest projection; for if l denote the angle which any plane having

to the following, that the sum of the moments of the

---

the sum of the projections upon it represented by a makes with that of the greatest projection, we shall have
$$a = \sqrt{(c^2 + c'^2 + c''^2)} . \cos . l,$$
therefore if l be invariable a is also.

If D represent the sum of the areas upon any plane, ϵ, ϵ', and ϵ'' the angles which a perpendicular to it at the origin of the co-ordinates makes with the axes z, y, and x respectively, and A, B, and C the sums of the projections of the areas upon the three planes xy, xz, and yz of the co-ordinates, then the following equation may be easily demonstrated.
$$D = A.\cos.\epsilon + B.\cos.\epsilon' + C.\cos.\epsilon''.$$
If the areas A, A', A'', &c. are supposed to be respectively double the triangles which have lines representing in direction and magnitude, the forces S, S', S'', &c. for their bases, and the origin of the co-ordinates of their points of application for their common vertex, then if N denote the moments of the forces projected upon the plane xy, it will represent also the sum of the projections of the areas A, A', A'', &c. upon the same plane. In like manner, if M and L represent the moments of the forces projected upon the planes xz and yz, they will also represent the projections of the areas A, A', A'', &c. upon these planes. It is therefore evident that the three quantities L, M, and N, and analogous quantities relative to the same system of forces, and to other planes, will have the same properties as the sums of the projections upon those planes.

In the above the origin of the co-ordinates, or the centre of moments, is supposed to be invariable, and the forces S, S', S'', &c. to be resolved in directions parallel to the respective planes, and to be moved parallel to themselves to these planes, and to act along them.

It therefore follows, that if the sums of the moments of the forces S, S', S'', &c. resolved along the three planes of the co-ordinates be known, the sum of the moments of the same forces resolved along any other plane passing through

forces which would make it turn about the three axes, be nothing relative to each of them *.

---

the centre of moments will be known from the following equation
$$D = N.\cos.\varepsilon + M.\cos.\varepsilon' + L.\cos.\varepsilon'',$$
in which D represents the sum sought, and ε, ε', and ε'' the angles which a perpendicular to any plane from the centre of the co-ordinates respectively makes with the axes of the co-ordinates z, y, and x.

The sum of the moments with respect to the plane of the greatest sum of the moments is represented by $\sqrt{L^2 + M^2 + N^2}$, and that of any plane making the ange l with it is equal to $\sqrt{L^2 + M^2 + N^2}.\cos.l$, if the angle l be a right angle then $\cos.l = 0$, and the sum of the moments relative to this plane vanishes.

If α, β, and γ represent the angles which the perpendicular to the plane of the greatest sum of the moments makes with the axes of the co-ordinates x, y, and z, its position will be determined by the following equations
$$\cos.\alpha = \frac{L}{\sqrt{L^2 + M^2 + N^2}},$$
$$\cos.\beta = \frac{M}{\sqrt{L^2 + M^2 + N^2}},$$
$$\cos.\gamma = \frac{N}{\sqrt{L^2 + M^2 + N^2}}.$$

If lines be taken upon the perpendiculars to each plane from the centre of moments proportional to the sums of the moments upon these planes, the line representing the greatest sum will be the diagonal of a parallelepiped constructed upon the lines representing the the three sums L, M and N.

The composition of moments, therefore follows the same laws as that of forces, the greatest sum and the perpendicular to its plane having place instead of the resultant and its direction.

* If the forces are all supposed parallel to each other and their directions to make the angles α, β, and γ with the

Let it be supposed that the bodies m, m', m'', &c. are only acted upon by the force of gravity: as its action

---

co-ordinates x, y and z respectively; the equations (n) may be changed into the following

$$0 = \cos.\alpha.\Sigma.mSy - \cos.\beta.\Sigma.mSx,$$
$$0 = \cos.\alpha.\Sigma.mSz - \cos.\gamma.\Sigma.mSx,$$
$$0 = \cos.\gamma.\Sigma.mSy - \cos.\beta.\Sigma.mSz,$$

the third of which is a consequence of the two first; but as by trigonometry $\cos.^2\alpha + \cos.^2\beta + \cos.^2\gamma = 1$, we may determine from these equations the angles α, β, and γ. By supposing, for abridgment,

$$\Sigma.mSx = mSx + m'S'x' + m''S''x'' + \&c. = L,$$
$$\Sigma.mSy = mSy + m'S'y' + m''S''y' + \&c. = M,$$
$$\Sigma.mSz = mSz + m'S'z' + m''S''z'' + \&c. = N,$$

the following equations will readily be found,

$$\cos.\alpha = \frac{L}{\sqrt{(L^2+M^2+N^2)}}, \quad \cos.\beta = \frac{M}{\sqrt{(L^2+M^2+N^2)}},$$
$$\cos.\gamma = \frac{N}{\sqrt{(L^2+M^2+N^2)}}.$$

The position of the bodies being given with respect to the three axes, it is necessary in order that all motion of rotation may be destroyed, that the system should be placed relative to the direction of the forces, in such a position as to cause the direction to make with the three axes the angles determined above.

If the quantities L, M, and N vanish, the angles α, β, and γ will remain indeterminate, and the system will be in equilibrio in any position. Therefore, if the sum of the products of parallel forces by their distances from three planes perpendicular to each other be nothing with respect to each of these planes, the effect of the forces to turn the body about the common point of intersection of these planes will be nothing.

is the same upon all the bodies, and as the directions of gravity may be conceived to be the same in all the extent of the system, we shall have

$$S = S' = S'' = \&c.;$$

$$\left(\frac{\delta s}{\delta x}\right) = \left(\frac{\delta s}{\delta x'}\right) = \left(\frac{\delta s}{\delta x''}\right) = \&c.;$$

$$\left(\frac{\delta s}{\delta y}\right) = \left(\frac{\delta s}{\delta y'}\right) = \left(\frac{\delta s}{\delta y''}\right) = \&c.;$$

$$\left(\frac{\delta s}{\delta z}\right) = \left(\frac{\delta s}{\delta z'}\right) = \left(\frac{\delta s}{\delta z''}\right) = \&c.;$$

the three equations *(n)* will be satisfied, whatever may be the direction of *s*, or of gravity, by means of the three following

$$0 = \Sigma.mx;\ 0 = \Sigma.my;\ 0 = \Sigma.mz.\quad (p)$$

The origin of the co-ordinates being supposed fixed, it will destroy parallel to the three axes, the forces

$$S.\left(\frac{\delta s}{\delta x}\right).\Sigma.m;\ S.\left(\frac{\delta s}{\delta y}\right).\Sigma.m;\ \text{and}\ S.\left(\frac{\delta s}{\delta z}\right).\Sigma.m$$

respectively; by composing these three forces, we shall have a resultant equal to $S.\Sigma.m$; that is, equal to the weight of the system.

This origin of the co-ordinates, about which we here suppose the system to be in equilibrio, is a very remarkable point in it, on this account, that being sustained, if the system is supposed to be only acted upon by the force of gravity, it will remain in equilibrio, whatever situation we may give it about this point; which is called the centre of gravity of the system. Its position is determined by the condition, that if we make any plane whatever pass by this point, the sum of the products of each body by its distance from the plane is nothing; for the distance is a linear function of the co-ordinates x, y and z of the body*; if it be therefore multiplied by the mass of the body, the sum of these products will be nothing in consequence of the equations *(o)*.

In order to fix the position of the centre of gravity, let X, Y, and Z be its three co-ordinates relative to a given point; let x, y and z be the co-ordinates of m relative to the same point, x' y', and z' those of m', and so on; the equations *(o)* will then give

$$0 = \Sigma.m.(x-X);$$

but we have $\Sigma.mX = X.\Sigma.m$, $\Sigma.m$ being the entire mass of the system; we shall have therefore

$$X = \frac{\Sigma.mx}{\Sigma.m}.$$

We shall have in like manner

$$Y = \frac{\Sigma.my}{\Sigma.m}; \quad Z = \frac{\Sigma.mz}{\Sigma.m}$$

but as the co-ordinates X, Y, and Z determine only one point, it is evident that the centre of gravity of a body is only one point.

The three preceeding equations give

$$X^2 + Y^2 + Z^2 = \frac{(\Sigma.mx)^2 + (\Sigma.my)^2 + (\Sigma.mz)^2}{(\Sigma.m)^2};$$

---

* Let $Ax' + By' + Cz' = 0$ be the equation to a plane passing through the centre of gravity, which is supposed to be the origin of the co-ordinates; then if x, y, and z are the co-ordinates of the body, its distance from that plane will be $\dfrac{Ax + By + Cz}{\sqrt{(A^2 + B^2 + C^2)}}$, which is a linear function of the co-ordinates x, y and z of the body.

which equation may be altered to this form*

$$x^2+y^2+z^2 = \frac{\Sigma.m.(x^2+y^2+z^2)}{\Sigma.m}$$

$$-\frac{\Sigma.mm'.\{(x'-x)^2+(y'-y)^2+(z'-z)^2\}}{(\Sigma.m)^2};$$

the finite integral $\Sigma.mm'.\{(x'-x)^2+(y'-y)^2+(z'-z)^2\}$ expresses the sum of all the products, similar to that which is contained under the characteristic Σ,

---

* In order to render this evident it will be sufficient to give an example, in which only three bodies m, m', and m'' are considered with respect to the co-ordinates of x. In this case $\frac{\Sigma.m(x^2)}{\Sigma.m}$ is equal to $\frac{mx^2+m'x'^2+m''x''^2}{m+m'+m''}$ and $\frac{\Sigma.mm'\{(x'-x)^2\}}{(\Sigma.m)^2}$ to

$$\frac{mm'x'^2-2mm'xx'+mm'x^2+mm''x''^2-2mm''x''x+mm''x^2+}{(\Sigma.m)^2}$$

$$\frac{m'm''x''^2-2m'm''x''x'+m'm''x'^2}{},$$ if both the numerator and denominator of the first quantity are multiplied by $\Sigma.m$ or $m+m'+m''$, it will become

$$\frac{m^2x^2+mm'x'^2+mm''x''^2+mm'x^2+m'^2x'^2+m'm''x''^2+mm''}{(\Sigma.m)^2}$$

$$x^2+m'm''x'^2+m''^2x''^2},$$

which, by subtracting the last quantity, gives

$$\frac{m^2x^2+2mm'xx'+m'^2x'^2+2m'm''x'x''+2mm''xx''+m''^2x''^2}{(\Sigma.m)^2}$$

$$\frac{(\Sigma.mx)^2}{(\Sigma.m)^2}.$$

which we are able to form, from considering by two and two all the bodies of the system. We shall thus have the distance of the centre of gravity from any fixed point whatever, by means of the distances of the bodies of the system from the same fixed point, and their mutual distances. By determining in this manner, the distances of the centre of gravity from any three fixed points whatever, we shall have its position in space, which is a new method of determining it*.

We have extended the denomination of centre of gravity to a point of any system whatever of bodies, either having or not having weight, determined by the three co-ordinates, X, Y, and Z†.

---

* As the last term of the second member of the equation is independent of the given point, if the values of the first term be determined with respect to three given points not in the same straight line taken either within or without the system, we shall have the distances of its centre of gravity from these points, and consequently its position with respect to them. If the bodies were in the same plane two points would have been sufficient, and if in the same line, one. If the given points be taken in the bodies of the system, the position of its centre of gravity will be given solely by the masses and their respective distances. This method of finding the centre of gravity is independent of the consideration of three planes.

† It is evident from the principle of virtual velocities, that the centre of gravity of a system of bodies connected together in any manner, is generally the highest or the lowest possible when the system is in equilibrio.

Let m, m', m'', &c. be the centres of gravity of a number of bodies connected together, whose weights are denoted by

16. It is easy to apply the preceding results to the
the powers S, S', S'', &c. acting respectively upon these centres, and let s, s', s'', &c. represent lines respectively drawn from them to any horizontal plane. If the position of the system be disturbed in an indefinitely small degree, we shall have, in the case of equilibrium, the equation of virtual velocities $S.\delta s + S'.\delta s' + S''\delta s'' + \&c. = 0$;
the quantity $Ss + S's' + S''s'' + \&c.$ is therefore either a maximum or a minimum. If the sum of the weights S, S', S'', &c. be represented by G, and the distance of the centre of gravity of the system from the horizontal plane by g, we shall have the following equation
$$S.s + S'.s' + S''.s'' + \&c. = G.g.$$
As the first member of this equation is either a maximum or a minimum, the second is also, consequently the distance of the centre of gravity from the horizontal plane is either a maximum or a minimum when the system is in a state of equilibrium.

When the distance of the centre of gravity from an horizontal plane is a maximum, the equilibrium of the system of heavy bodies is unstable, and if moved in an indefinitely small degree would not return to its former state; on the contrary, when the distance of the centre of gravity is a minimum the system if moved from the state of equilibrium, would, after oscillating some time, return to it. This may be exemplified in the case of a cylinder with an elliptical base, which, when placed upon an horizontal plane with the edge of contact in the line passing along an extremity of the major axis, will have the distance of its centre of gravity from the plane a maximum and its position unstable, and the contrary when placed with the edge of contact in a line passing through an extremity of the minor axis. The above are the only positions in which there can be an equilibrium.

equilibrium of a solid body having any figure whatever, by supposing it formed of an indefinite number of points invariably connected with each other. Let dm represent one of these points, or an indefinitely small molecule of the body, and let $x, y,$ and z be the rectangular co-ordinates of this molecule; again, let $P, Q,$ and R be the forces by which it is actuated parallel to the axes of $x, y,$ and z; the equations (m) and (n) of the preceding No. will be changed into the following;

$$0 = \int P.dm\ ;\ \ 0 = \int Q.dm\ ;\ \ 0 = \int R.dm\ ;$$
$$0 = \int(Py - Qx)dm\ ;\ \ 0 = \int(Pz - Rx)dm\ ;$$
$$0 = \int(Ry - Qz)dm\ ;$$

the integral sign $\int$ is relative to the molecule dm, and ought to be extended to the whole mass of the solid*.

If the body could only turn about the origin of the co-ordinates, the three last equations would be sufficient for its equilibrium.

---

* It is easy to perceive that in the case of a solid body, which may be supposed to be composed of an indefinitely great number of points invariably connected together, the quantity $\Sigma.mS\left(\frac{\delta s}{\delta x}\right)$ becomes $\int P.dm$; for $S\left(\frac{\delta s}{\delta x}\right)$ is equivalent to P, and $\int dm$ to $\Sigma.m$; in like manner $\Sigma.mS\left\{y.\left(\frac{\delta s}{\delta x}\right) - x.\left(\frac{\delta s}{\delta y}\right)\right\}$ becomes $\int(Py - Qx)dm$, for $\Sigma.S\left(\frac{\delta s}{\delta x}\right).y\ m$ is equivalent to $\int Py.dm$, and $\Sigma.S\left(\frac{\delta s}{\delta y}\right).x\ m$ to $\int Qx.dm$.

CHAP. IV.

Of the equilibrium of fluids.

17. To have the laws of the equilibrium and of the motion of each of the fluid molecules, it is necessary to know their figure, which is impossible; but we have no occasion to determine these laws except for the fluids considered in a mass, and then the knowledge of the forms of their molecules becomes useless. Whatever may be these figures and the dispositions which result in the integral molecules; all the fluids taken in the mass ought to offer the same phenomena in their equilibrium, and in their motions, so that the observation of these phenomena does not enable us to learn any thing respecting the configuration of the fluid molecules. These general phenomena are founded upon the perfect mobility of the molecules, which are thus able to give way to the slightest effort. This mobility is the characteristic property of fluids: it distinguishes them from solid bodies, and serves to define them. From hence it results, that for the equilibrium of a fluid mass, each molecule ought to be in equilibrio, in con-

sequence of the forces which solicit it, and the pressure which it sustains from the surrounding molecules. Let us develope the equations which result from this property.

In order to accomplish it, we will consider a system of fluid molecules forming an indefinitely small rectangular parallelepiped. Let x, y, and z be the three rectanglar co-ordinates of the angle of this parallelepiped the nearest to the origin of the co-ordinates. Let dx, dy, and dz be the three dimensions of this parallelepiped; let p represent the mean of all the pressures which the different points of the side dy, dz of the parallelepiped that is nearest the origin of the co-ordinates experiences; and p' the same quantity relative to the opposite side. The parallelepiped, in consequence of the pressure which acts upon it, will be solicited parallel to the axis of x by a force equal to $(p-p').dy.dz$; $p'-p$ is the differential of p taken by making x alone to vary; for although the pressure of p' acts in a different direction to that of p, nevertheless the pressure which a point of the fluid experiences being the same in all directions; $p'-p$ may be considered as the difference of two forces indefinitely near and acting in the same direction.; we shall therefore have $p'-p = \left(\frac{dp}{dx}\right).dx$; and $(p-p')$ $dy.dz = -\left(\frac{dp}{dx}\right).dx.dy.dz$*.

---

* In (fig. 14) let AX, AY, and AZ represent the axes of x, y, and z respectively, and ah a molecule of the fluid in the form of a rectangular parallelepiped, whose fa-

Let P, Q, and R be three accelerating forces which also act upon the fluid molecule, parallel to the respective axes of x, y, and z: if the density of the parallelepiped is named ρ, its mass will be $\rho.dx.dy.dz$, and the product of the force P by this mass, will be the entire resulting force which moves it; this mass will consequently be solicited parallel to the axis of x, by the force $\left\{\rho P - \left(\dfrac{dp}{dx}\right)\right\}.dx.dy.dz$. It will in like manner be solicited parallel to the axes of y and z, by the forces $\left\{\rho Q - \left(\dfrac{dp}{dy}\right)\right\}.dx.dy.dz$, and $\left\{\rho R - \left(\dfrac{dp}{dz}\right)\right\}.dx.dy.dz$; we shall therefore have, in consequence of the equation (b) of No. 3,

$$0 = \left\{\rho P - \left(\dfrac{dp}{dx}\right)\right\}.\delta x + \left\{\rho Q - \left(\dfrac{dp}{dy}\right)\right\}.\delta y + \left\{\rho R - \left(\dfrac{dp}{dz}\right)\right\}.\delta z,$$

---

ces bh, ag, and ad are respectively parallel to the planes YAX, ZAX, and YAZ. Suppose that the co-ordinates of the angular point b of the molecule are x, y, and z, and that $bg = dx$, $bd = dy$, and $ba = dz$; also, let mo represent the quantity and direction of the mean of all the forces acting upon the face dy, dz of the parallelogram, or the force p, and nq the mean of all the forces acting upon the opposite face fg of the molecule, or the force p'.

In this case p is supposed to be a function of the co-ordinates x, y, and z, consequently for the opposite side of the parallelepiped to that formed by dy and dz, as x becomes $x + dx$, the pressure p is changed into $p + \left(\dfrac{dp}{dx}\right).dx$, which

or $\delta p = \rho.\{P.\delta x + Q.\delta y + R.\delta z\}$*.

The second member of this equation ought to be an exact variation like the first; which gives the following equations of partial differentials,

$$\left(\frac{d.\rho P}{dy}\right) = \left(\frac{d.\rho Q}{dx}\right); \quad \left(\frac{d.\rho P}{dz}\right) = \left(\frac{d.\rho R}{dx}\right);$$
$$\left(\frac{d.\rho Q}{dz}\right) = \left(\frac{d.\rho R}{dy}\right);$$

from which we may obtain

$$0 = P.\left(\frac{dQ}{dz}\right) - Q.\left(\frac{dP}{dz}\right) + R.\left(\frac{dP}{dy}\right) - P.\left(\frac{dR}{dy}\right) +$$

---

for that side must necessarily act in an opposite direction to p. It therefore follows that $\left(\frac{dp}{dx}\right)dx$, the difference of the two pressures, when multiplied by dy and dz, gives the whole force arising from pressure that acts upon the parallelepiped in the direction of the co-ordinates, which should be taken negatively, as it tends to diminish them, and, in the case of equilibrium, must be equal to the moving force $\rho P.dx.dy.dz$.that acts in an opposite direction.

* The equation

$$0 = \left\{\rho P - \left(\frac{dp}{dx}\right)\right\}.\delta x + \left\{\rho Q - \left(\frac{dp}{dy}\right)\right\}.\delta y + \left\{\rho R - \left(\frac{dp}{dz}\right)\right\}.\delta z,$$

by transposition, becomes

$$\left(\frac{dp}{dx}\right)\delta x + \left(\frac{dp}{dy}\right)\delta y + \left(\frac{dp}{dz}\right)\delta z = \rho\{P.\delta x + Q.\delta y + R.\delta z\},$$

but $\left(\frac{dp}{dx}\right)\delta x + \left(\frac{dp}{dy}\right)\delta y + \left(\frac{dp}{dz}\right)\delta z$ is equivalent to $\left(\frac{\delta p}{\delta x}\right)\delta x + \left(\frac{\delta p}{\delta y}\right)\delta y + \left(\frac{\delta p}{\delta z}\right)\delta z$ or δp, page 45, therefore $\delta p = \rho(P.\delta x + Q.\delta y + R.\delta z)$.

$$Q.\left(\frac{dR}{dx}\right) - R.\left(\frac{dQ}{dx}\right)\dagger.$$

---

† The proof that if $\rho\{P.\delta x + Q.\delta y + R.\delta z\}$ is an exact differential, the equations
$$\left(\frac{d.\rho P}{dy}\right) = \left(\frac{d.\rho Q}{dx}\right), \quad \left(\frac{d.\rho P}{dz}\right) = \left(\frac{d.\rho R}{dx}\right)$$
$$\left(\frac{d.\rho Q}{dz}\right) = \left(\frac{d.\rho R}{dy}\right),$$
have place, may be given as follows. Suppose $u = f(x,y)$, then $du = pdx + qdy$ is an exact differential, if $p = \left(\frac{du}{dx}\right)$ and $q = \left(\frac{du}{dy}\right)$; by differentiating y alone in the first, and x alone in the second of the two last equations, we shall have
$$\left(\frac{dp}{dy}\right) = \left(\frac{d^2 u}{dxdy}\right) \text{ and } \left(\frac{dq}{dx}\right) = \left(\frac{d^2 u}{dydx}\right),$$
but $\left(\frac{d^2 u}{dxdy}\right) = \left(\frac{d^2 u}{dydx}\right)$, therefore $\left(\frac{dp}{dy}\right) = \left(\frac{dq}{dx}\right)$.
In like manner, if $u = f(x,y,z)$, by differentiation $du = pdx + qdy + rdz$, in which equation $p = \left(\frac{du}{dx}\right)$, $q = \left(\frac{du}{dy}\right)$ and $r = \left(\frac{du}{dz}\right)$; let z be supposed constant, then $du = pdx + qdy$, which gives $\left(\frac{dp}{dz}\right) = \left(\frac{dq}{dx}\right)$; also if y and x are supposed alternately constant, the resulting equations will respectively give $\left(\frac{dp}{dz}\right) = \left(\frac{dr}{dx}\right)$ and $\left(\frac{dq}{dz}\right) = \left(\frac{dr}{dy}\right)$. In the above ρP may be substituted for p, ρQ for q, and ρR for r.

By differentiating the equations $\left(\frac{d.\rho P}{dy}\right) = \left(\frac{d.\rho Q}{dx}\right)$, $\left(\frac{d.\rho P}{dz}\right) = \left(\frac{d.\rho R}{dx}\right)$, $\left(\frac{d.\rho Q}{dz}\right) = \left(\frac{d.\rho R}{dy}\right)$, and afterwards multiplying the first by R, the second by $-Q$, and the third by P; they will give by adding together, as ρ disappears, the

This equation expresses the relation which should exist amongst the forces P, Q and R, that the equilibrium may be possible*.

---

following equation $0 = P \cdot \left(\dfrac{dQ}{dz}\right) - Q \cdot \left(\dfrac{dP}{dz}\right) + R \cdot \left(\dfrac{dP}{dy}\right)$
$- P \cdot \left(\dfrac{dR}{dy}\right) + Q \cdot \left(\dfrac{dR}{dx}\right) - R \cdot \left(\dfrac{dQ}{dx}\right).$

* Suppose an incompressible fluid not acted upon by the force of gravity to be contained in a vessel that has a number of cylinders attached to the sides of it, to which a number of moveable pistons are adapted. Let the areas or bases of the cylinders or pistons be represented by A, A', A'', &c. also suppose S, S', S'', &c. to denote the powers applied to the pistons having the bases A, A', A'', &c. respectively, and that these powers, which act upon each other by the intervention of the fluid, are in equilibrio. Let p represent, in this case, the pressure upon the area denoted by unity of the surface of the vessel or the base of the pistons, then pA, pA', pA'', &c. will denote the respective pressures of the fluid upon the bases of the pistons, but these pressures are equal to the forces which act upon the pistons, therefore $S = pA$, $S = pA'$, $S = pA''$, &c. Let a part of the pistons be pushed downwards, then it is evident that the other part of the pistons must be elevated by an equal quantity of water to that depressed, so that if δs, $\delta s'$, $\delta s''$, &c. represent the depressions or the elevations of the respective pistons whose bases are A, A', A'', &c. we shall have the equation

$$A \cdot \delta s + A' \cdot \delta s' + A'' \cdot \delta s'' + \&c. = 0,$$

regarding the spaces through which the pistons were depressed as positive, and the spaces through which they were elevated as negative. Let this equation be multiplied by p, then

$$pA \cdot \delta s + p \cdot A' \cdot \delta s' + pA'' \cdot \delta s'' + \&c. = 0,$$

or by substitution

$$S \cdot \delta s + S' \cdot \delta s' + S'' \cdot \delta s'' + \&c. = 0;$$

which is the equation of virtual velocities.

If the fluid be free at its surface, or in certain parts

Suppose that T, T', T'', &c. represent the different powers which act upon a molecule whose co-ordinates are x, y, and z; these powers being directed to certain fixed centres, the distance of which from the molecule solicited are respectively t, t', t'', &c.

Let the co-ordinates of these fixed centres referred to the origin of the co-ordinates x, y, and z, and respectively parallel to them be a, b, c; a', b', c', &c. we shall then have

$$P = \frac{T}{t}(x-a) + \frac{T'}{t'}(x-a') + \frac{T''}{t''}(x-a'') + \&c.$$

$$Q = \frac{T}{t}(y-b) + \frac{T'}{t'}(y-b') + \frac{T''}{t''}(y-b'') + \&c.$$

$$R = \frac{T}{t}(z-c) + \frac{T'}{t'}(z-c') + \frac{T''}{t''}(z-c'') + \&c.$$

$$t = \sqrt{\{(x-a)^2 + (y-b)^2 + (z-c)^2\}},$$
$$t' = \sqrt{\{(x-a')^2 + (y-b')^2 + (z-c')^2\}},$$
&c. &c.

As the equilibrium is possible, when the fluid molecules are solicited by forces directed towards fixed centres, which are functions of the distances of the points of application from these centres; we may substitute the above values of P, Q, and R, in the equation

$$\frac{\delta p}{\rho} = (P.\delta x + Q.\delta y + R.\delta z),$$

which then becomes

$$\frac{\delta p}{\rho} = \frac{T}{t}\{(x-a)\delta x + (y-b)\delta y + (z-c)\delta z\}$$
$$+ \frac{T'}{t'}\{(x-a')\delta x + (y-b')\delta y + (z-c')\delta z\}$$
$$+ \frac{T''}{t''}\{(x-a'')\delta x + (y-b'')\delta y + (z-c'')\delta z\}$$
$$+ \&c.$$

and is equivalent to

$$\frac{\delta p}{\rho} = T.\delta t + T'.\delta t' + T''.\delta t'' + \&c.$$

of its surface, the value of p will be nothing in those parts; we shall therefore have $\delta p = 0$*, provided that we subject the variations δx, δy, and, δz to appertain

---

The sum $\Sigma(\delta p)$ taken throughout the whole extent of any indefinitely narrow canal, which either re-enters into itself, or is terminated at two points of the exterior surface of the fluid mass, is always nothing; on the supposition that the resistance of the sides, if the fluid be contained in a vessel is regarded, and that the canal is imagined in this case to have one of its extremities terminated at a point of its side. It may therefore be concluded, that for all the cases of the equilibrium of a fluid, the following equation has place throughout the whole extent of the mass,
$$\Sigma\{\rho(T.\delta t + T'.\delta t' + T''.\delta t'' + \&c.)\} = 0.$$
In this equation the products of ρT, $\rho T'$, &c. are proportional to the moving forces with which each power acts upon the molecule. Let S, S', S'', &c. represent the moving forces which are the resultants of the powers which respectively act upon each fluid molecule, and s, s', s'', &c. the lines drawn respectively in the directions of the forces S, S', S'', &c. from each fluid molecule; then the above equation is equivalent to the following
$$S.\delta s + S'.\delta s' + S''.\delta s'' + \&c. = 0.$$
This equation is similar to those deduced from the principal of virtual velocities for the equilibrium of a point or a system.

* This will be the case not only when $p=0$, but likewise when p is a constant quantity, which also gives $\delta p = 0$. For instance when the atmosphere presses equally upon the surface of the fluid.

to this surface: by fulfilling these conditions we shall consequently have

$$0 = P.\delta x + Q.\delta y + R.\delta z^*.$$

---

* Suppose for example, a fluid mass to be acted upon by a force S tending to the centre of the co-ordinates, and let one of its molecules be placed at the distance r from that centre, having x, y, and z, for its rectangular co-ordinates, then $r = \sqrt{x^2 + y^2 + z^2}$: the force S resolved parallel to these co-ordinates gives $\frac{Sx}{r}, \frac{Sy}{r}$, and $\frac{Sz}{r}$ for the forces in their respective directions. These forces, taken negatively as they tend to diminish the co-ordinates, should be substituted for their respective values P, Q, and R, in the preceding equations. When they are substituted in the equation $0 = P.\delta x + Q.\delta y + R.\delta z$ they will give, by the suppression of the common factor $-\frac{S}{r}$,

$$0 = x.\delta x + y.\delta y + z.\delta z,$$

which is an exact differential, therefore the equilibrium is possible. This equation when integrated becomes $x^2 + y^2 + z^2 = c^2$, which is the equation to a sphere, consequently, the fluid will assume a spherical form. If r is very great the surface of the fluid may be regarded as a plane, as is the case with the surface of a fluid in equilibrio in a vessel when only acted upon by the force of gravity.

Let the force S be supposed to vary as the nth power of its distance from the centre of the co-ordinates, and to be represented by Ar^n, also, let p represent the pressure upon an area of the surface denoted by unity, then the equation $p = \varrho \int \{P.\delta x + Q.\delta y + R.\delta z\}$ will, by a proper substitution, be changed into the following

$$p = A\varrho \int r^{n-1}(x\delta x + y\delta y + z\delta z),$$

but $x\delta x + y\delta y + z\delta z = r\delta r$, therefore

If $\delta u = 0$ be the differential equation of the surface, we shall have $P.\delta x + Q.\delta y + R.\delta z = \lambda.\delta u$, λ being a function of x, y, and z; from which it follows, by No. 3, that

$$p = A\varrho \int r^n \delta r = \frac{A\varrho}{n+1} \cdot (x^2 + y^2 + z^2)^{\frac{n+1}{2}} + \text{const.}$$

This is the value of the pressure referred to unity of the surface which acts upon the molecule that has x, y, and z for its co-ordinates.

The equation of equilibrium may be used to find the form which a fluid retains when it has an uniform rotatory motion round a fixed axis, by adding the centrifugal force to the given accelerating forces which act upon the molecules. Let the axis of z be that of rotation, n the angular velocity common to all the points of the fluid mass, and $r = \sqrt{x^2 + y^2}$ the distance of any point of it from the axis of rotation; then, as the centrifugal force of the point is equal to the square of its velocity divided by its distance from the axis of z, it will be represented by $n^2 r$, which when multiplied by the variation of its direction gives $n^2 r \delta r = n^2 x \delta x + n^2 y \delta y$. If this value be added to the formula $P.\delta x + Q.\delta y + R.\delta z$, it will not prevent it from being an exact differential, for the centrifugal force of a point may be considered as a force of repulsion, the intensity of which is a function of the distance of the point from the axis of rotation: we shall therefore have the equation

$$(P + n^2 x)\delta x + (Q + n^2 y)\delta y + R.\delta z = 0$$

for the differential equation of the surface of the laminæ and of the free surface of the fluid. That the velocity n may be uniform it is requisite that the forces P, Q, and R should arise from the mutual attraction of the molecules, or from attractions in the directions of lines joining the molecules and the axis of rotation, or from forces acting towards points which have the same motion as the fluid mass.

the resultant of the forces P, Q, and R, ought to be perpendicular to those parts of the surface where the fluid is free.

Let us suppose that the variation $P.\delta x + Q.\delta y + R.\delta z$ is exact; which circumstance has place by No. 2, when the forces P, Q, and R are the results of attractive forces.

Naming this variation $\delta\varphi$, we shall have $\delta p = \rho.\delta\varphi$; ρ therefore should be a function of p and of φ, and as by integrating this differential equation, we have φ a function of p; we shall have p a function of ρ. The

---

By way of example, let us find the form of the surface of a quantity of water contained in a cylindrical vessel open at the top, having a rotatory motion round its axis which is vertical. As water is an incompressible and homogeneous fluid, the above equation will be sufficient, let therefore the origin of the co-ordinates be at the centre of the base of the vessel having z for the vertical axis of rotation, and let g represent the force of gravity, then we shall have $P=0$, $Q=0$ and $R=-g$, consequently the equation of the surface of the fluid becomes by substitution

$$n^2 x \delta x + n^2 y \delta y - g \delta z = 0$$

which by integration gives

$$\frac{n^2}{2}(x^2 + y^2) - gz = \text{const.}$$

In this case it is evident from the equation that the upper part of the fluid will form the surface of a paraboloid of which the solid content is given, as it will be equal to one half that of the water contained in the vessel. The equation of the generating parabola is $y^2 = \frac{2g}{n^2} z$ as appears from making x to vanish.

pressure p is therefore the same for all the molecules of which the density is the same; therefore δp is nothing relative to the surfaces of the laminæ of the fluid mass in which the density is constant; and we shall have by relation to these surfaces

$$0 = P.\delta x + Q.\delta y + R.\delta z.$$

It therefore follows, that the resultant of the forces which act upon each fluid molecule, is in the state of equilibrium perpendicular to the surfaces of these laminæ; which have been named on that account (couches de niveau) laminæ of level. This condition is always fulfilled if the fluid is homogeneous and incompressible; because in this case the laminæ to which this resultant is perpendicular are all of the same density*.

---

* In the case of the equilibrium of an elastic fluid, the pressure is found by experiment to vary as the density, consequently p may be supposed equal to $a\varrho$. If $p = a\varrho$ then $\varrho = \frac{p}{a}$, let this value of ϱ be substituted in the equation $dp = \varrho d\varphi$ and it will give $adp = pd\varphi$, consequently $d.\log.p = \frac{1}{a}.d\varphi$. If the fluid be homogeneal and of the same temperature, a is constant and the equation possible. By integration we shall have $\log.p = \frac{\varphi}{a} + \log.c$, or $p = c.e^{\frac{\varphi}{a}}$, c being a constant quantity, and e the base of the system of logarithms which has unity for its modulus. As this value of p and consequently that of $\varrho = \frac{p}{a}$ are functions of the sole variable φ, the

For the equilibrium of an homogeneous fluid mass, the exterior surface of which is free, and covers a solid mass fixed and of any figure whatever; it is therefore

---

pressure and the density will be the same for all the extent of each lamina of level as in the case of heterogeneal fluids; the densities of the different laminæ are determined by the equation

$$\varrho = \frac{p}{a} = \frac{c}{a} \cdot e^{\frac{\varphi}{a}}.$$

If in an homogeneal fluid the temperature be not the same throughout the mass, the quantity a will not be constant; let t denote the temperature, then a will be a function of t, but a, if variable, must be a function of φ, consequently t must be a function of φ; it is therefore necessary in the case of equilibrium that the temperature be uniform throughout each lamina of level, as well as the pressure and the density, which are likewise functions of φ. The temperature may vary according to any law in passing from one lamina to another, but this law being given the laws of the pressure and the density will be determined by the following equations,

$$p = c \cdot e^{\int \frac{d\varphi}{a}}, \quad \varrho = \frac{c}{a} \cdot e^{\int \frac{d\varphi}{a}};$$

c being a constant quantity.

In the case of the equilibrium of the atmospheric air, let it be supposed that a small vertical cylindrical column of it is continued from the surface of the earth to an indefinite height, which must be supposed in equilibrio independent of the surrounding air that may be conceived to become immoveable. The force of gravity can without sensible error be supposed to act in the direction of the cylinder along its

necessary and it is sufficient, first, that the differential $P.\delta x + Q.\delta y + R.\delta z$ be exact; secondly, that the resultant of the forces at the exterior surface be directed perpendicularly to this surface.

---

whole extent, and in the case of equilibrium, the density, the pressure, and the temperature may be considered as invariable throughout the whole mass of an indefinitely thin horizontal lamina. Let z represent the distance of any lamina whatever from the surface of the earth, ϱ its density, r the radius of the earth, g the force of gravity at its surface, $g' = \dfrac{g r^2}{(r+z)^2}$ the force of gravity at that altitude, and p its elastic force, ϱ, g' and p being functions of z, dz the breadth of the lamina and a the area of its base. Then if p be the pressure upon a portion of the surface represented by unity, that upon the higher part of the lamina will be Ap and that upon the lower $A(p+dp)$, but the excess Adp of the first pressure above the second is equal to the weight of the lamina or $A\varrho g dz$, therefore $\delta p = -\varrho g' \delta z$. This equation may be obtained by proper substitution from the general equation
$$\delta p = \varrho (P.\delta x + Q.\delta y + R.\delta z),$$
in which P and Q will vanish and R be equal to $-g'\delta z$.

CHAP. V.

General principles of motion of a system of bodies.

18. W E have in No. 7 reduced the laws of the motion of a point to those of its equilibrium, by resolving the instantaneous motion into two others, of which one remains, and the other is destroyed by the forces which solicit the point: the equilibrium between these forces and the motion lost by the body, has given us the differential equations of its motion. We now proceed to make use of the same method to determine the motion of a system of bodies m, m', m'', &c. Let therefore mP, mQ, and mR, be the forces which solicit m parallel to the axes of its respective rectangular co-ordinates x, y, and z; let $m'P'$, $m'Q'$, and $m'R'$ be the forces which solicit m' parallel to the same axes, and in a similar manner with the rest; and let the time be represented by t. The partial forces $m \cdot \dfrac{dx}{dt}$, $m \cdot \dfrac{dy}{dt}$, and $m \cdot \dfrac{dz}{dt}$ of the body m at any instant whatever will become

in the following

$$m.\frac{dx}{dt} + m.d.\frac{dx}{dt} - m.d.\frac{dx}{dt} + mP.dt;$$

$$m.\frac{dy}{dt} + m.d.\frac{dy}{dt} - m.d.\frac{dy}{dt} + mQ.dt;$$

$$m.\frac{dz}{dt} + m.d.\frac{dz}{dt} - m.d.\frac{dz}{dt} + mR.dt;$$

and as the sole forces

$$m.\frac{dx}{dt} + m.d.\frac{dx}{dt}; \quad m.\frac{dy}{dt} + m.d.\frac{dy}{dt}; \quad m.\frac{dz}{dt} + m.d.\frac{dz}{dt};$$

remain; the forces

$$-m.d.\frac{dx}{dt} + mP.dt; \quad -m.d.\frac{dy}{dt} + mQ.dt;$$

$$-m.d.\frac{dx}{dt} + mR.dt;$$

are destroyed*.

---

* The forces which are destroyed during the motion of the system at any instant, will evidently form an equation of equilibrium for it at that instant. If in this equation of equilibrium the bodies undergo an indefinitely small change in their position, the moments of the forces according to the principle of virtual velocities will be equal to nothing; the the forces destroyed are mP, mQ, mR, $m'P'$, &c. $-m.\frac{d^2x}{dt^2}$, $-m.\frac{d^2y}{dt^2}$, $-m.\frac{d^2z}{dt^2}$, $-m'.\frac{d^2x'}{dt^2}$, &c. whose moments are $mP.\delta x$, $mQ.\delta y$, $mR.\delta z$, &c. $-m.\frac{d^2x}{dt^2}.\delta x$, $-m.\frac{d^2y}{dt^2}\delta y$, $-m.\frac{d^2z}{dt^2}\delta z$. &c. the general formula of equilibrium is therefore, when multiplied by -1, as follows

$$m\left(\frac{d^2x}{dt^2} - P\right).\delta x + m\left(\frac{d^2y}{dt^2} - Q\right).\delta y + m\left(\frac{d^2z}{dt^2} - R\right).\delta z +$$

By distinguishing in these two expressions, the letters m, x, y, z, P, Q, and R successively by one, two, &c. marks, we shall have the forces destroyed in the bodies m', m'', &c. This being premised; if we multiply these forces respectively by the variations δx, δy, δz, $\delta x'$; &c. of their directions; the principle of virtual velocities explained in No. 14, will give, by supposing dt constant, the following equation;

$$\left.\begin{array}{l} 0 = m.\delta x.\left\{\dfrac{d^2x}{dt^2} - P\right\} + m.\delta y.\left\{\dfrac{d^2y}{dt^2} - Q\right\} \\ + m.\delta z.\left\{\dfrac{d^2z}{dt^2} - R\right\} + m'.\delta x'.\left\{\dfrac{d^2x'}{dt^2} - P'\right\} \\ + m'.\delta y'.\left\{\dfrac{d^2y}{dt^2} - Q'\right\} + m'.\delta z'.\left\{\dfrac{d^2z'}{dt^2} - R'\right\} \\ \&c^* . \end{array}\right\}; (P)$$

$$m'\left(\dfrac{d^2x'}{dt^2} - P'\right).\delta x' + \&c. = 0.$$

In the equation of equilibrium of the forces destroyed, in order that they may be equal to nothing, either the forces $m.\dfrac{d^2x}{dt^2}$, $m.\dfrac{d^2y}{dt^2}$, &c. or the forces mP, mQ, &c. must be taken negatively, although in the motion of the system they may tend to increase the co-ordinates.

* The expression $d^2x.\delta x + d^2y.\delta y + d^2z.\delta z$ is independent of the position of the axes of the co-ordinates x, y, and z; as may be proved in the following manner.

Let the rectangular co-ordinates $x_{\prime}$, $y_{\prime}$, and $z_{\prime}$ be substituted for those above mentioned, having the same origin but referred to other axes; then it may easily be demonstrated that

$$x = \alpha x_{\prime} + \beta y_{\prime} + \gamma z_{\prime}$$
$$y = \alpha' x_{\prime} + \beta' y_{\prime} + \gamma' z_{\prime}$$
$$z = \alpha'' x_{\prime} + \beta'' y_{\prime} + \gamma'' z_{\prime}$$

We can extract from this equation by means of the particular conditions of the system, as many of the

---

the co-efficients α, β, γ, α', &c. being constant quantities, and only dependent upon the respective positions of the two systems of co-ordinates. The co-ordinates x, y, and z are relative to the same points as the co-ordinates $x_{\prime}$, $y_{\prime}$, and $z_{\prime}$, consequently $x^2+y^2+z^2=x_{\prime}^2+y_{\prime}^2+z_{\prime}^2$; this expression gives the six following conditional equations,

$\alpha^2+\alpha'^2+\alpha''^2=1$, $\beta^2+\beta'^2+\beta''^2=1$, $\gamma^2+\gamma'^2+\gamma''^2=1$, $\alpha\beta+\alpha'\beta'+\alpha''\beta''=0$, $\alpha\gamma+\alpha'\gamma'+\alpha''\gamma''=0$, $\beta\gamma+\beta'\gamma'+\beta''\gamma''=0$,

from which it appears, that three of the nine co-efficients are indeterminate quantities.

If the expressions of x, y, and z are twice differentiated, they will become

$$d^2x=\alpha d^2x_{\prime}+\beta d^2y_{\prime}+\gamma d^2z_{\prime},$$
$$d^2y=\alpha' d^2x_{\prime}+\beta' d^2y_{\prime}+\gamma' d^2z_{\prime},$$
$$d^2z=\alpha'' d^2x_{\prime}+\beta'' d^2y_{\prime}+\gamma'' d^2z_{\prime},$$

the following variations may likewise be obtained

$$\delta x=\alpha\delta x_{\prime}+\beta\delta y_{\prime}+\gamma\delta z_{\prime},$$
$$\delta y=\alpha'\delta x_{\prime}+\beta'\delta y_{\prime}+\gamma'\delta z_{\prime},$$
$$\delta z=\alpha''\delta x_{\prime}+\beta''\delta y_{\prime}+\gamma''\delta z_{\prime}.$$

By substituting these values and regarding the equations of condition between α, β, γ, α', &c. we shall find that

$$d^2x\delta x+d^2y\delta y+d^2z\delta z=d^2x_{\prime}\delta x_{\prime}+d^2y_{\prime}\delta y_{\prime}+d^2z_{\prime}\delta z_{\prime}.$$

If the same substitutions are made in the expressions for the right lined distances between the different bodies of the system represented by f, f', f'', &c. the quantities α, β, γ, α', &c. will equally disappear, and the transformed expressions will retain the same form. Thus x, y, and z being the co-ordinates of the body m, and x', y', and z' those of the body m', their distance f will be equal to $\sqrt{(x'-x)^2+(y'-y)^2+(z'-z)^2}$. If the axes are changed, the first co-ordinates will become $x_{\prime}$, $y_{\prime}$, and $z_{\prime}$, and the second $x_{\prime}'$, $y_{\prime}'$,

variations as we have conditions; by afterwards equalling separately to nothing the remaining co-efficients of

---

and $z_{\prime}'$, also
$$x' = \alpha x_{\prime}' + \beta y_{\prime}' + \gamma z_{\prime}',$$
$$y' = \alpha' x_{\prime}' + \beta' y_{\prime}' + \gamma' z_{\prime}',$$
$$z' = \alpha'' x_{\prime}' + \beta'' y_{\prime}' + \gamma'' z_{\prime}'.$$

By substitution and having regard to the before mentioned equations of condition, we shall have $f = \sqrt{(x_{\prime}' - x_{\prime})^2 + (y_{\prime}' - y_{\prime})^2 + (z_{\prime}' - z_{\prime})^2}$; a similar proof may be given for the quantities f', f'', &c.

It follows from the above, that if the system is only acted by forces which are proportional to some functions of the distances f, f', f'', &c. between the bodies; and the equations of condition of the system solely depend upon the mutual situation of the bodies or the lines f, f', f'', &c. the general formula of dynamic will be the same for the transformed co-ordinates $x_{\prime}$, $y_{\prime}$, and $z_{\prime}$, as for the original ones x, y, and z. If, therefore, the different values of x, y, and z for each body are found with respect to the time by integration of the different equations deduced from this formula, and those values are taken $x_{\prime}$, $y_{\prime}$, and $z_{\prime}$, we shall have these more general values for x, y, and z.
$$x = \alpha x_{\prime} + \beta y_{\prime} + \gamma z_{\prime},$$
$$y = \alpha' x_{\prime} + \beta' y_{\prime} + \gamma' z_{\prime},$$
$$z = \alpha'' x_{\prime} + \beta'' y_{\prime} + \gamma'' z_{\prime},$$
in which the nine co-efficients α, β, γ, α', &c. contain three indeterminate quantities, as there are only six conditional equations amongst them.

If the values of $x_{\prime}$, $y_{\prime}$, and $z_{\prime}$ contain all the constant quantities necessary to complete the different integrals; the three indeterminate quantities will be mixed with some of the other six constant quantities, they will also make up those that

the variations, we shall have all the necessary equations for determining the motions of the different bodies of the system*.

19. The equation *(P)* contains many general principles of motion, which we shall proceed to develope. The variations δx, δy, δz, $\delta x'$, &c. will be evidently

---

are wanting, without which the solution would be incomplete. Thus by means of these three new constant quantities which may be introduced after the calculation, it will be possible to suppose the same number of the other constant quantities equal to nothing, or to some determinate quantities; which will often much facilitate and simplify the calculation. Vide the Mechanique Analytique of Lagrange.

* Although the effects of the forces of impulsion or percussion may be calculated in the same manner as those of accelerating forces, yet when the whole impressed velocity only is required, its successive increments can be neglected, and the forces of impulsion considered in what follows as equivalent to the impressed motions.

Let therefore S, S', S'', &c. represent the forces of impulsion applied to any body m of the system in the directions of the lines s, s', s'', &c. and suppose that the velocity given to this body may be resolved into three velocities, represented by $\dot{x}$, $\dot{y}$, and $\dot{z}$ in the direction of the co-ordinates x, y, and z, we shall have by changing the accelerating forces $\frac{d^2 x}{dt^2}$, $\frac{d^2 y}{dt^2}$, and $\frac{d^2 z}{dt^2}$ into the velocities $\dot{x}$, $\dot{y}$, and $\dot{z}$, the general equation

$$\Sigma . m(\dot{x}\delta x + \dot{y}\delta y + \dot{z}\delta z) - \Sigma(S.\delta s + S'.\delta s' + S''.\delta s'') + \&c. = 0.$$

This equation will give as many particular ones, as we shall have independent variations, after they have been reduced to the smallest number possible by means of the conditional equations belonging to the system.

subjected to all the conditions of the connection of the parts of the system, if they are supposed equal to the differentials dx, dy, dz, dx', &c.* This supposition is

---

If the system is continuous and of an invariable figure as a solid body, or variable as flexible bodies and fluids; by denoting its whole mass by m and any one of its molecules by dm, it may be considered as an assemblage or system of an indefinitely great number of molecules; each represented by dm and acted upon by the accelerating forces S, S', S'', &c.; and it will be sufficient in the general equation to substitute dm for m, and for the sign Σ, S or the sign of integration relative to the whole extent of the body, that is, to the instantaneous position of all its molecules, but independent of the successive positions of each molecule. For a fuller detail respecting solid bodies, I refer the reader to the seventh chapter of this work.

* It is necessary in this case that the equations of condition should not contain the time t, which sometimes happens, as for instance, if one of the bodies be forced to move upon a surface which is itself moving according to some given law, there will then be an equation of condition of the co-ordinates and the time t, for the equation of the surface at any instant, which may be represented as follows,
$$F(t, x, y, z, x', y', z', \&c.) = 0.$$
In the equation of equilibrium of a system formed by those forces which are supposed equal to nothing when it is in motion, it is necessary, in order that the indefinitely small change in its position according to the principle of virtual velocities may be proper, that the co-ordinates of the bodies in the new position of the system when substituted should satisfy that equation. These co-ordinates are $x+\delta x, y+\delta y$, and $z+\delta z$ for the body m, $x'+\delta x'$, $y'+\delta y'$, and $z'+\delta z'$ for the body m', &c. which should satisfy the above conditional

therefore permitted, consequently the equation (P) will give by integration

$$\Sigma . m . \frac{dx^2 + dy^2 + dz^2}{dt^2} = c + 2 . \Sigma . \int m \, (P.dx + Q.dy + R.dz) ; \quad (Q)$$

c being a constant quantity introduced by the integration.

If the forces P, Q, and R are the results of attracting forces directed towards fixed points, and of attracting

---

equation when respectively substituted for x, y, z, x', y', z', &c.; the differential of the function F,

$$\left(\frac{dF}{dx}\right).\delta x + \left(\frac{dF}{dy}\right).\delta y + \left(\frac{dF}{dz}\right).\delta z + \left(\frac{dF}{dx'}\right).\delta x' + \&c.$$

will then be equal to nothing, t being regarded as constant and the variations of the co-ordinates x, y, z; x', y', z'; &c. denoted by the characteristic δ. But as the co-ordinates of the bodies are functions of the time, the complete differentiation of F with respect to t, x, y, z, x', &c. being regarded as functions of t, will be equal to nothing. We shall therefore have the following equation,

$$T.dt + \left(\frac{dF}{dx}\right)dx + \left(\frac{dF}{dy}\right)dy + \left(\frac{dF}{dz}\right)dz + \left(\frac{dF}{dx'}\right)dx' + \&c. = 0.$$

$T.dt$ being the differential of F taken with respect to the time which is contained explicitly in this function. If $T.dt$ be equal to nothing it is evident that the former equation will coincide with this, by taking $\delta x = dx$, $\delta y = dy$, &c.

From the above it appears, that when the time is not explicitly contained in any of the equations of condition, the virtual velocities of the moving bodies along the axes of their co-ordinates may be supposed equal to the differentials of these co-ordinates, or the spaces passed over by their projections upon these axes during the time dt.

forces of the bodies one towards another, the function $\Sigma.fm.(P.dx+Q.dy+R.dz)$ is an exact integral. In fact, the parts of this function relative to attracting forces directed towards fixed points, are by No. 8, exact integrals. This is equally true with respect to the parts which depend upon the mutual attractions of the bodies of the system; for if the distance between m and m' is called f, and the attraction of m' for m, $m'F$, the part of $m.(P.dx+Q.dy+R.dz)$ relative to the attraction of m' for m, will be, by the above cited No. equal to $-mm'.Fdf$, the differential df being taken by only making the co-ordinates x, y, and z to vary*.

---

* As $m'F$ is the accelerating force of m arising from the attraction of m' which acts along the line f, its components in the directions of the axes of x, y, and z are $m'F.\dfrac{x'-x}{f}$, $m'F.\dfrac{y'-y}{f}$, and $m'F.\dfrac{z'-z}{f}$, we shall therefore have the following equation with respect to this force alone

$$P.dx+Q.dy+R.dz=\frac{m'F}{f}\{(x'-x)dx+(y'-y)dy+(z'-z)dz\}.$$

In a similar manner, P', Q', and R' denote the components of the accelerating forces which act upon m' parallel to the same axes, we shall have relative to the force mF the equation

$$P'dx'+Q'dy'+R'dz'=\frac{mF}{f}.\{(x-x')dx+(y-y')dy'+(z-z')dz'\}.$$

If, after having multiplied the first of these equations by m and the second by m', we add them together, they will introduce into the expression $\Sigma m(Pdx+Qdy+Rdz)$ the term

But the re-action being equal and contrary to the action, the part of $m'.(P'.dx'+Q'.dy'+R'.dz')$ relative to the attraction of m to m', is equal to $-mm'.Fdf$, if the co-ordinates x', y', and z' alone vary in f; the part of the function $\Sigma.m(Pdx+Qdy+Rdz)$ relative to the reciprocal attraction of m and of m', is therefore $-mm'.Fdf$; the whole being supposed to vary in f. This quantity is an exact differential when f is a function of F, or when the attraction varies as a function of the distance, which we shall always suppose; the function $\Sigma.m.(Pdx+Qdy+Rdz)$ is therefore an exact differential, whenever the forces which act upon the bodies of the system, are the result of their mutual attraction or of attracting forces directed towards certain fixed points. Let $d\varphi$ represent this differential and v the velocity of m, v' that of m', &c.; we shall then have

$$\Sigma.mv^2 = c + 2\varphi. \quad (R)$$

This equation is analogous to the equation (g) of No. 8; it is the analytical traduction of the principle of the preservation of the living forces. The product of the mass of a body by the square of its velocity, is called

---

$$\frac{mm'F}{f}.\{(x'-x)(dx-dx')+(y'-y)(dy'-dy)+(z'-z)(dz'-dz)\}.$$

If the equation $f^2=(x-x')^2+(y-y')^2+(z-z')^2$ be completely differentiated it will give

$$fdf=(x-x')(dx-dx')+(y-y')(dy-dy')+(z-z')(dz-dz');$$

the preceding term by substitution therefore becomes $-mm'Fdf$.

its living force. The principle upon which we are treating consists in this, that the sum of the living forces, or the whole living force of the system is constant, if the system be not solicited by any forces; and if the bodies be solicited by any forces whatever, the sum of the increments of the whole living force is the same, whatever may be the curves described by each of the bodies, provided, that their points of departure and arrival are the same.

This principle has place only in the cases in which the motions of the bodies change by insensible gradations. If these motions experience sudden changes, the living force is diminished by a quantity which we shall determine in the following manner. The analysis which has conducted us to the equation (P) of the preceding number, gives in this case instead of that equation, the following*,

---

* The equation $\Sigma.m.\left\{\left(\dfrac{d^2x}{dt^2}-P\right).\delta x+\left(\dfrac{d^2y}{dt^2}-Q\right).\delta y+\left(\dfrac{d^2z}{dt^2}-R\right).\delta z\right\}$ of No. 19 is equivalent to $\Sigma.m.\left(\dfrac{d^2x}{dt^2}.\delta x+\dfrac{d^2y}{dt^2}.\delta y+\dfrac{d^2z}{dt^2}.\delta z\right)-\Sigma.m.(P.\delta x+Q.\delta y+R.\delta z)$, or $\Sigma.m.\left(\dfrac{\delta x}{dt}.d.\dfrac{dx}{dt}+\dfrac{\delta y}{dt}.d.\dfrac{dy}{dt}+\dfrac{\delta z}{dt}.d.\dfrac{dz}{dt}\right)-\Sigma.m.(P.\delta x+Q.\delta y+R.\delta z)$;

which equation if the differences $\Delta.\dfrac{dx}{dt}$, $\Delta.\dfrac{dy}{dt}$, and $\Delta.\dfrac{dz}{dt}$ are substituted for the differentials $d.\dfrac{dx}{dt}$, $d.\dfrac{dy}{dt}$, and $d.\dfrac{dz}{dt}$, is changed into the following,

$0=\Sigma.m.\left(\dfrac{\delta x}{dt}.\Delta.\dfrac{dx}{dt}+\dfrac{\delta y}{dt}.\Delta.\dfrac{dy}{dt}+\dfrac{\delta z}{dt}.\Delta.\dfrac{dz}{dt}\right)-\Sigma.m.(P.\delta x+Q.\delta y+R.\delta z)$.

$$0 = \Sigma . m . \left\{ \frac{\partial x}{dt} . \Delta . \frac{dx}{dt} + \frac{\partial y}{dt} . \Delta . \frac{dy}{dt} + \frac{\partial z}{dt} . \Delta . \frac{dz}{dt} \right\} - \Sigma . m . (P . \delta x + Q . \delta y + R . \delta z);$$

$\Delta . \frac{dx}{dt}$, $\Delta . \frac{dy}{dt}$, and $\Delta . \frac{dz}{dt}$ being the differentials of $\frac{dx}{dt}$, $\frac{dy}{dt}$, and $\frac{dz}{dt}$ from one instant to another; which differentials become finite when the motions of the bodies receive finite alterations in an instant. We may suppose in this equation

$$\delta x = dx + \Delta . dx; \quad \delta y = dy + \Delta . dy; \quad \delta z = dz + \Delta . dz;$$

because the values of dx, dy, and dz are changed in the following instant into $dx + \Delta . dx$, $dy + \Delta . dy$, and $dz + \Delta . dz$, these values of dx, dy, and dz satisfy the conditions of the connection of the parts of the system; we shall therefore have

$$0 = \Sigma . m . \left\{ \left(\frac{dx}{dt} + \Delta . \frac{dx}{dt} \right) . \Delta . \frac{dx}{dt} + \left(\frac{dy}{dt} + \Delta . \frac{dy}{dt} \right) . \Delta . \frac{dy}{dt} + \left(\frac{dz}{dt} + \Delta . \frac{dz}{dt} \right) . \Delta . \frac{dz}{dt} \right\} - \Sigma . m . \{ P . (dx + \Delta . dx) + Q . (dy + \Delta . dy) + R . (dz + \Delta . dz) \}.$$

This equation should be integrated as an equation of finite differences relative to the time t, of which the variations are indefinitely small, as well as the variations of x, y, z, x', &c. Let us denote by $\Sigma_{,}$ the finite integrals resulting from this integration, to distinguish them from the preceding finite integrals, relative to all the bodies of the system. The integral of $mP . (dx + \Delta . dx)$, is evidently the same as $\int mP . dx$; we shall therefore have*

---

* The integral of $mP . (dx + \Delta . dx)$ is $\int m . P . dx$, for $dx + \Delta . dx = \delta x$, but $\int m . P . \delta x$ is equivalent to $\int m . P . dx$, therefore

$$\text{constant} = \Sigma.m.\frac{dx^2+dy^2+dz^2}{dt^2} + \Sigma_{,}.\Sigma.m.\left\{\left(\Delta.\frac{dx}{dt}\right)^2 + \left(\Delta.\frac{dy}{dt}\right)^2 + \left(\Delta.\frac{dz}{dt}\right)^2\right\} - 2\Sigma.\int.m.(P.dx+Q.dy+R.dz);$$

if v, v', v'', &c. denote the velocities of m, m', m'', &c. we shall have

$$\Sigma.mv^2 = \text{const.} - \Sigma_{,}.\Sigma.m.\left\{\left(\Delta.\frac{dx}{dt}\right)^2 + \left(\Delta.\frac{dy}{dt}\right)^2 + \left(\Delta.\frac{dz}{dt}\right)^2\right\} + 2\Sigma.\int.m.(P.dx+Q.dy+R.dz).$$

The quantity contained under the sign Σ, being necessarily positive, we may perceive that the living force of the system is diminished by the mutual action of the bodies, whenever during the motion some of the variations $\Delta.\frac{dx}{dt}$, $\Delta.\frac{dy}{dt}$, &c. are finite. The preceding equation moreover offers a simple means for determining this diminution.

At each sudden variation in the motion of the system, it is possible to suppose the velocity of m resolved into

$\Sigma_{,}mP.(dx+\Delta dx)$ is equivalent to $\int mP.dx$. In this equation the integral of $\frac{dx}{dt}.\Delta.\frac{dx}{dt}$ is supposed to be $\frac{1}{2}.\frac{dx^2}{dt^2}$.

Those who wish to obtain information respecting finite differences and their integrals, may find it in the two works of S. F. Lacroix, upon the Calcul Differentiel and Calcul Integral, in the Traite De Calcul Differentiel et Calcul Integral, par J. A. J. Cousin, and in the 19th Leçon of the Leçons sur Le Calcul Des Fonctions, par J. L. Lagrange, &c. &c.

two others, one of which v remains during the following instant; the other of which V is destroyed by the action of the other bodies; but the velocity of m being $\frac{\sqrt{(dx^2+dy^2+dz^2)}}{dt}$ before this resolution, and changing afterwards into $\frac{\sqrt{\{(dx+\Delta.dx)^2+(dy+\Delta.dy)^2+(dz+\Delta.dz)^2\}}}{dt}$, it is easy to perceive that we have

$$V^2 = \left(\Delta.\frac{dx}{dt}\right)^2 + \left(\Delta.\frac{dy}{dt}\right)^2 + \left(\Delta.\frac{dz}{dt}\right)^2 ;$$

the preceding equation ought therefore to be put into this form
$\Sigma.mv^2 =$ const. $- \Sigma_{,}.\Sigma.mV^2 + 2\Sigma.\int.m.(P.dx + Q.dy + R.dz^*$.

---

* The equation $\Sigma.mv^2 = c + 2\varphi$, when differentiated with respect to t, becomes $\Sigma.mv.\frac{dv}{dt} = \frac{d\varphi}{dt}$, or $mv.\frac{dv}{dt} + m'v'.\frac{dv'}{dt} + mv''.\frac{dv''}{dt} + \&c. = S.\frac{ds}{dt} + S'.\frac{ds'}{dt} + S''.\frac{ds''}{dt} + \&c.$ which equation has place for a system of bodies connected with each other in any manner whatever, which reciprocally attract or repel each other, or are attracted towards or repelled from fixed centres by any forces S, S', S'', &c.; naming the mutual distances of the bodies which attract or repel each other, or their distances from fixed centres of attraction or repulsion s, s', s'', &c.; taking the quantities S, S', S'', &c. which represent the forces, positively or negatively, according as these forces are repulsive or attractive; as the first tend to increase and the second to decrease the distances s, s', s'', &c. This principle also has place in the movement of inelastic fluids so long as they form a continuous mass and there is no impact amongst their molecules.

20. If in the equation (P) of No. 18, we suppose
$$\delta x' = \delta x + \delta x_{\prime}' ; \quad \delta y' = \delta y + \delta y_{\prime}' ; \quad \delta z' = \delta z + \delta z_{\prime}' ;$$

If the forces S, S', S'', &c. are respectively functions of the distances s, s', s'', &c. along which they act, which may always be supposed when these forces are independent of each other; or in general if the quantities S, S', S'', &c. are such functions of s, s', s'', &c. that the quantity $S . \frac{ds}{dt} + S' . \frac{ds'}{dt} + S'' . \frac{ds''}{dt} + $ &c. is the differential of a function of s, s', s'', &c. which can be denoted by $F(s, s', s'',$ &c.$)$, the integral of the above equation will be
$$mv^2 + m'v'^2 + m''v''^2 + \&c. = c + 2F(s, s', s'', \&c.),$$
c being a constant quantity. In this case the forces S, S', S'', &c. which act along the lines s, s', s'', &c. will be represented by $\frac{d.F(s, s', s'', \&c.)}{ds}$, $\frac{d.F(s, s', s'', \&c.)}{ds'}$, &c. respectively.

Let a, a', a'', &c. be the respective values of s, s', s'', &c. and V, V', V'', &c. the respective velocities of m, m', m'', &c. at a given instant; the preceding equation referred to this same instant will give
$$mV^2 + m'V'^2 + m''V''^2 + \&c. = c + 2F(a, a', a'', \&c.) ;$$
consequently $c = mV^2 + m'V'^2 + m''V''^2 + \&c. - 2F(a, a', a'',$ &c.$)$, therefore, by substitution, we have the following general equation,
$$mv^2 + m'v'^2 + m''v''^2 + \&c. = mV^2 + m'V'^2 + m''V''^2 + \&c. + 2F(s, s', s'', \&c.) - 2F(a, a', a'', \&c.).$$

This is the general equation of the preservation of the living forces, from which it is evident, that the whole living force of the system depends upon the active forces, such as the forces of attraction or repulsion, or springs, &c. ; and upon the position of the bodies relative to the centre of

$$\delta x'' = \delta x + \delta x_{\prime}''; \quad \delta y'' = \delta y + \delta y_{\prime}''; \quad \delta z'' = \delta z + \delta z_{\prime}'';$$
&c.

---

these forces; therefore if at two instants the bodies are at the same distances from these centres, the sum of their living forces will be the same.

If the bodies should strike each other, or meet with obstacles which cause a sudden alteration in their motions, the above formula may be applied to the bodies during these alterations, however short they may be; thus denoting by V, V', V'', &c. their velocities at the commencement of the sudden change, and by v, v', v'', &c. their velocities at its termination, also by a, a', a'', &c. the values of the distances s, s', s'', &c. at the beginning, and by A, A', A'', &c. their values at the end of the same action, the following equation will have place,

$mV^2 + m'V'^2 + m''V''^2 + $ &c. $- mv^2 - m'v'^2 - m''v''^2 - $ &c. $= 2F(a, a', a'',$ &c.$) - 2F(A, A', A'',$ &c.$).$ Which shews that the difference of the living forces at the commencement, and at the end of the action will be $2F(a, a', a'',$ &c.$) - 2F(A, A', A'',$ &c.$).$ This expression may have any finite value whatever, however small the difference between the respective quantities a, a', a'', &c. and A, A', A'', &c.

When perfectly elastic bodies strike each other, either directly or by the intervention of levers or any machines whatever, the compression and the restitution of the shapes of the bodies follow the same law, and the action is supposed to continue until the bodies are restored unto the same respective positions in which they were when the compression commenced. In this case we shall have $a = A$, $a' = A'$, $a'' = A''$, &c. and consequently $F(a, a', a'',$ &c.$) = F(A, A', A'',$ &c.$)$; therefore the living force will be the same after as before the shock.

The following proof in the case of two elastic bodies impinging upon each other, is derived from the laws of the

by substituting these variations in the expressions of the variations δf, $\delta f'$, $\delta f''$, &c. of the mutual distances

---

motion of elastic bodies. Let V and V' be the velocities of two elastic bodies m and m' before the shock, v and v' their velocities after, also let u represent their common velocity at the time of contact; then as may be seen in the elementary treatises upon Mechanics

$v = 2u - V$, $v' = 2u - V'$, $u = \dfrac{mV + m'V'}{m + m'}$, and the quantity $mv^2 + m'v'^2$ by substitution becomes $m(2u - V)^2 + m'(2u - V')^2 = 4u^2(m + m') - 4u(mV + m'V') + mV^2 + m'V'^2 = mV^2 + m'V'^2$.

In the shock of in-elastic bodies, the action is only supposed to continue, until the bodies have acquired the velocities which hinder their acting upon each other any longer. As therefore the effect of these velocities upon the mutual action of the bodies is nothing, if we had impressed these same velocities before the action it would have been the same, in consequence of the velocities composed of these, and of the velocities properly belonging to the bodies.

Again therefore, it would be the same if the velocities impressed were equal and directly contrary to those above mentioned; for the action will not be varied by supposing that these impressed velocities were destroyed by the opposite velocities. It consequently follows, that in the shock of hard bodies the velocities v, v', v'', &c. after the shock, ought to be such, that if we give these same velocities to the bodies m, m', m'', &c. in a contrary direction, the equation $mV^2 + m'V'^2 + $ &c. $- mv^2 - m'v'^2 - m''v''^2 - $ &c. $= 2F(a, a', a'', \&c.) - 2F(A, A', A'', \&c.)$ given above will equally have place. But the terms which compose the second member, as they depend upon the mutual action of the bodies, will necessarily remain the same;

of the bodies of the system, of which we have given the values in No. 15; we shall find that the variations

---

therefore the value of $mV^2 + m'V'^2 + m''V''^2 + $ &c. $- mv^2 - m'v'^2 - m''v''^2 - $&c. will not be changed by composing the velocities V, V', V'', &c. and the velocities v, v', v'', &c. with the velocities $-v$, $-v'$, $-v''$, $-$&c. respectively. If therefore the velocity composed of V and $-v$ is represented by M, the velocity composed of V' and $-v'$ by M', &c. the following equation will have place,
$mV^2 + m'V'^2 + m''V''^2 + $&c.$- mv^2 - m'v'^2 - m''v''^2 - $&c.$= m M^2 + m'M'^2 + m''M''^2 + $&c.
as the velocities $v-v$, $v'-v'$, $v''-v''$, &c. vanish.

Because V, V', V'', &c. are the velocities before the shock, and v, v', v'', &c. the velocities after the same, it is evident that M, M', M'', &c. will be the velocities lost by the shock; therefore $mM^2 + m'M'^2 + m''M''^2 + $&c. will be the living force which results from these velocities, consequently this conclusion may be obtained. That in the shock of hard bodies, there is a loss of living forces equal to the living force which the same bodies would have had, if each of them should have been actuated by the velocity which it lost by the shock. Vide the Principes Fondamentaux de L'Equilibre et du Movement, par L. M. N. Carnot; and the Theorie des Fonctions Analytiques, par J. L. Lagrange.

It is evident from what has been said, that when the bodies of a system move in a resisting medium, or are subjected to friction from fixed obstacles, the living forces are constantly diminished and would at length entirely cease if the bodies should not be kept in motion by other forces. The formula $\Sigma . \int m(P.dx + Q.dy + R.dz)$ in these cases would not be an exact integral.

In the equation given in the notes at page 144, we may suppose that the variations δx, δy, and δz are proportional

δx, δy, and δz will disappear from these expressions.*
If the system be free, that is if it have no one of its parts connected with foreign bodies; the conditions relative to the mutual connection of the bodies depending only upon their mutual distances, the variations δx, δy, and δz will be independent of these conditions; from which it follows, that by substituting for $\delta x'$, $\delta y'$, $\delta z'$, $\delta x''$, &c. their preceding values in the equation (P), we ought to equal separately to nothing the co-

---

to the velocities $\dot{x}$, $\dot{y}$, and $\dot{z}$ which the bodies have received by impulsion. We shall then have the equation

$$\Sigma.\{m(\dot{x}^2+\dot{y}^2+\dot{z}^2)-(P\dot{x}+Q\dot{y}+R\dot{z})\}=0,$$

in which $\Sigma m(\dot{x}^2+\dot{y}^2+\dot{z}^2)$ represents the whole living force of the system.

* At No. 15, $f=\sqrt{\{(x'-x)^2+(y'-y)^2+(z'-z)^2\}}$.
$f'=\sqrt{\{(x''-x)^2+(y''-y)^2+(z''-z)^2\}}$,
&c. &c.

consequently by differentiation

$$\delta f=\frac{(x'-x)(\delta x'-\delta x)+(y'-y)(\delta y'-\delta y)+(z'-z)(\delta z'-\delta z)}{\sqrt{\{(x'-x)^2+(y'-y)^2+(z'-z)^2\}}}.$$
&c. &c.

If $\delta x+\delta x_{,}'$ be substituted for $\delta x'$, $\delta y+\delta y_{,}'$ for δy, and $\delta z+\delta z_{,}'$ for δz, in the preceding value of δf, it will become

$$\delta f=\frac{(x'-x)(\delta x+\delta x_{,}'-\delta x)+(y'-y)(\delta y+\delta y_{,}'-\delta y)+(z'-z)(\delta z+\delta z_{,}'-\delta z)}{\sqrt{\{(x'-x)^2+(y'-y)^2+(z'-z)^2\}}}$$

$$=\frac{(x'-x)\delta x_{,}'+(y'-y)\delta y_{,}'+(z'-z)\delta z_{,}'}{\sqrt{\{(x'-x)^2+(y'-y)^2+(z'-z)^2\}}};$$

which does not contain either δx, δy, or δz. In like manner by making the proper substitutions, the same quantities will disappear from the values of $\delta f'$, $\delta f''$, &c.

efficients of the variations δx, δy, and δz, which gives these three equations,

$$0 = \Sigma . m . \left(\frac{d^2 x}{dt^2} - P\right); \quad 0 = \Sigma . m . \left(\frac{d^2 y}{dt^2} - Q\right);$$

$$0 = \Sigma . m . \left(\frac{d^2 z}{dt^2} - R\right).$$

Let us suppose that X, Y, and Z are the three co-ordinates of the centre of gravity of the system; we shall have by No. 15,

$$X = \frac{\Sigma . m x}{\Sigma . m}; \quad Y = \frac{\Sigma . m y}{\Sigma . m}; \quad Z = \frac{\Sigma . m z}{\Sigma . m};$$

consequently the following equations may be obtained,

$$0 = \frac{d^2 X}{dt^2} - \frac{\Sigma . m P}{\Sigma . m}; \quad 0 = \frac{d^2 Y}{dt^2} - \frac{\Sigma . m Q}{\Sigma . m};$$

$$0 = \frac{d^2 Z}{dt^2} - \frac{\Sigma . m R}{\Sigma . m} *;$$

the centre of gravity of the system therefore moves in the same manner, as if, all the bodies m, m', &c. were united at this centre, and all the forces which solicit the system applied to it.

If the system be only submitted to the mutual action

---

* If $0 = \Sigma . m . \left(\frac{d^2 x}{dt^2} - P\right)$, then $\Sigma . m . \frac{d^2 x}{dt^2} = \Sigma . m P$, but $X = \frac{\Sigma . m x}{\Sigma . m}$, consequently $\frac{d^2 X}{dt^2} = \frac{\Sigma . m . \frac{d^2 x}{dt^2}}{\Sigma . m}$, therefore $\frac{d^2 X}{dt^2} = \frac{\Sigma . m P}{\Sigma . m}$; in like manner it may be proved that $\frac{d^2 Y}{dt^2} = \frac{\Sigma . m Q}{\Sigma . m}$, and $\frac{d^2 Z}{dt^2} = \frac{\Sigma . m R}{\Sigma . m}$.

of the bodies which compose it, and to their reciprocal attractions, we shall have

$$0 = \Sigma.mP\ ;\quad 0 = \Sigma.mQ\ ;\quad 0 = \Sigma.mR\ ;$$

for, from expressing by p the reciprocal action of m and m' whatever may be its nature, and denoting by f the mutual distance of these two bodies, we shall have in consequence of this sole action

$$mP = \frac{p.(x-x')}{f}*\ ;\quad mQ = \frac{p.(y-y')}{f}\ ;\quad mR = \frac{p.(z-z')}{f}\ ;$$

$$m'P' = \frac{p(x'-x)}{f}\ ;\quad m'Q' = \frac{p.(y'-y)}{f}\ ;\quad m'R' = \frac{p.(z'-z)}{f}\ ;$$

from which we may obtain

$$0 = mP + m'P'\dagger\ ;\quad 0 = mQ + m'Q'\ ;\quad 0 = mR + m'R'\ ;$$

and it is evident that these equations have place also in the case in which the bodies exercise upon each other a finite action in an instant. Their reciprocal action will therefore make the integrals $\Sigma.mP$, $\Sigma.mQ$, and $\Sigma.mR$ disappear, consequently they are nothing

---

* If from one extremity of the line f a line be drawn parallel to the axis of x, and from the other extremity another perpendicular to it, we shall have by No. 1. notes, $f : x-x' :: p : \frac{p(x-x')}{f}$, or the resolved force acting in a direction parallel to the axis of x, therefore $mP = \frac{p(x-x')}{f}$; in like manner $mQ = \frac{p(y-y')}{f}$, $mR = \frac{p(z-z')}{f}$. &c.

† $mP + m'P' = \frac{p(x-x')}{f} + \frac{p(x'-x)}{f} = \frac{p(x-x'+x'-x)}{f} = 0$, &c.

when the system is not solicited by forces unconnected with it. In this case we have

$$0 = \frac{d^2 X}{dt^2}; \quad 0 = \frac{d^2 Y}{dt^2}; \quad 0 = \frac{d^2 Z}{dt^2};$$

and by integrating

$$X = a + bt; \quad Y = a' + b't; \quad Z = a'' + b''t;$$

a, b, a', b', a'', b'', being constant quantities. By extracting the time t, we shall have an equation of the first order either between X and Y, or between X and Z; from which it follows, that the motion of the centre of gravity is rectilinear. Moreover as its velocity is equal to $\sqrt{\left(\frac{dX}{dt}\right)^2 + \left(\frac{dY}{dt}\right)^2 + \left(\frac{dZ}{dt}\right)^2}$, or to $\sqrt{(b^2 + b'^2 + b''^2)}$, it is constant, and the motion is uniform.

It is evident from the preceding analysis, that this inalterability of the motion of the centre of gravity of a system of bodies, whatever may be their mutual actions, subsists also in the case in which some of the bodies lose during an instant by this action, a finite quantity of motion*.

---

* If the equations $X = \frac{\Sigma.mx}{\Sigma.m}$, $Y = \frac{\Sigma.my}{\Sigma.m}$, and $Z = \frac{\Sigma.mz}{\Sigma.m}$ are differentiated with respect to the time they will give the following

$$\Sigma.m.\frac{dX}{dt} = \Sigma.m.\frac{dx}{dt}, \quad \Sigma.m.\frac{dY}{dt} = \Sigma.m.\frac{dy}{dt}, \quad \Sigma.m.\frac{dZ}{dt} = \Sigma.m.\frac{dz}{dt}.$$

As the simultaneous impact of a part of the bodies will change in general the velocities of all the bodies on account of their mutual connection, let a, b, and c represent the velocities of the body m in the respective directions of the co-ordinates, or the values of $\frac{dx}{dt}$, $\frac{dy}{dt}$, and $\frac{dz}{dt}$ immediately

21. If we make

$$\delta x' = \frac{y'.\delta x}{y} + \delta x_{\prime}'; \quad \delta x'' = \frac{y''.\delta x}{y} + \delta x_{\prime}''; \quad \&c.$$

$$\delta y = \frac{-x.\delta x}{y} + \delta y_{\prime}; \quad \delta y' = \frac{-x'.\delta x}{y} + \delta y_{\prime}';$$

$$\delta y'' = \frac{-x''.\delta x}{y} + \delta y_{\prime}''; \quad \&c.$$

the variation δx will again disappear from the expres-

---

before the impact a', b', and c' those of m' &c. and A, B, and C the values of these velocities immediately after for the body m, A', B', and C' the values for the body m', &c. in this case before the impact we shall have

$$\Sigma m.\frac{dX}{dt} = \Sigma ma, \quad \Sigma m.\frac{dY}{dt} = \Sigma mb, \quad \Sigma m.\frac{dZ}{dt} = \Sigma mc,$$

and after

$$\Sigma m.\frac{dX}{dt} = \Sigma mA, \quad \Sigma m.\frac{dY}{dt} = \Sigma mB, \quad \Sigma m.\frac{dZ}{dt} = \Sigma mC.$$

The quantities of motion lost by all the bodies at the instant of impact should be such as would cause an equilibrium in the system; these forces in the respective directions of the co-ordinates x, y, and z are $ma - mA$, $mb - mB$, and $mc - mC$, which for all the points of the system are $\Sigma.ma - \Sigma.mA$, $\Sigma.mb - \Sigma.mB$, &c. These quantities, as the system is not supposed to contain any fixed point, are respectively equal to nothing, consequently $\Sigma.ma = \Sigma.mA$, $\Sigma.mb = \Sigma.mB$, &c. therefore the values of $\frac{dX}{dt}$, $\frac{dY}{dt}$, &c. are the same after as before the impact, and the velocity of the centre of gravity of the system is the same in quantity and direction.

The following is a proof of its truth in the simple case of the impact of two non-elastic bodies obtained in a very different manner.

sions δf, $\delta f'$, $\delta f''$, &c.; by supposing therefore the system free, as the conditions relative to the connection

---

Take upon the right line which the bodies describe any point whatever for the origin of the spaces described, let x and x' represent the distances of two non-elastic bodies m and m' respectively from that point at the end of the time t, and X that of their centre of gravity, then by the known laws of the motion of non-elastic bodies given in elementary treatises upon Mechanics $X = \dfrac{mx + m'x'}{m + m'}$; if v and v' be the respective velocities of the bodies m and m' before, and V their common velocity or the velocity of their centre of gravity after the shock then $V = \dfrac{mv + m'v'}{m + m'}$, as is well known; but the velocity of their centre of gravity before the shock is $\dfrac{dX}{dt} = \dfrac{m \cdot \dfrac{dx}{dt} + m' \cdot \dfrac{dx'}{dt}}{m + m'}$, therefore as $\dfrac{dx}{dt} = v$, and $\dfrac{dx'}{dt} = v'$, the velocity of their centre of gravity is the same before as after the shock.

Let the bodies which compose a free system be supposed to be acted upon only by impulses, and in the equation given in the notes page 144, let $\delta x + \delta x_{\prime}'$, $\delta y + \delta y_{\prime}'$, &c. be substituted for $\delta x'$, $\delta y'$, &c. and the forces S, S', S'', &c. be reduced to the rectangular forces P, Q, R, P', &c. then the following equations may easily be proved from what has preceded,

$$0 = \Sigma.(m\dot{x} - P), \quad 0 = \Sigma.(m\dot{y} - Q), \quad 0 = \Sigma.(m\dot{z} - R).$$

If the co-ordinates x, y, and z are referred to the centre of gravity of the system we shall have

$$X\Sigma.m = \Sigma.mx, \quad Y\Sigma.m = \Sigma.my, \quad Z\Sigma.m = \Sigma.mz,$$

which being differentiated relative to t, by making $dX = \dot{X}dt$, $dY = \dot{Y}dt$, $dZ = \dot{Z}dt$; $dx = \dot{x}dt$, $dy = \dot{y}dt$, $dz = \dot{z}dt$, $dx' = \dot{x}'dt$, &c. we shall have

of the parts of the system, only influence the variations δf, $\delta f'$, &c. the variation δx is independent and arbitrary; therefore if we substitute in the equation (P) of No. 18, in the places of $\delta x'$, $\delta x''$, &c.; δy, $\delta y'$, $\delta y''$, &c. their preceding values; we ought to equal separately to nothing the co-efficient of δx, which gives

$$0 = \Sigma . m . \frac{x d^2 y - y d^2 x}{dt^2} + \Sigma . m . (Py - Qx);$$

from which we shall obtain by integrating with respect to the time t,

$$c = \Sigma . m . \frac{x dy - y dx}{dt} + \Sigma . \int . m . (Py - Qx) . dt;$$

c being a constant quantity.

We may in this integral change the co-ordinates y, y', &c. into z, z', &c. provided that we substitute instead of the forces Q, Q', &c. parallel to the axis of y, the forces R, R', &c. parallel to the axis of z; which gives

$$c' = \Sigma . m . \frac{x dz - z dx}{dt} + \Sigma . \int . m . (Pz - Rx) . dt;$$

c' being a new constant quantity. We shall have in like manner

$$c'' = \Sigma . m . \frac{y dz - z dy}{dt} + \Sigma . \int . m . (Qz - Ry) . dt;$$

c'' being a third constant quantity.

$$\dot{X} \Sigma m = \Sigma \dot{x} m, \quad \dot{Y} \Sigma m = \Sigma \dot{y} m, \quad \dot{Z} \Sigma m = \Sigma \dot{z} m,$$

and consequently

$$\ddot{X} \Sigma m - \Sigma P = 0, \quad \ddot{Y} \Sigma m - \Sigma Q = 0, \quad \ddot{Z} \Sigma m - \Sigma R = 0.$$

These equations shew that the velocities given to the centre of gravity are the same as would be given if all the bodies of the system were united at it and received at the same time the impulses ΣP, ΣQ, and ΣR.

Let us suppose that the bodies of the system are only submitted to their mutual action, and to a force directed towards the origin of the co-ordinates. If we name, as above, p the reciprocal action of m and of m', we shall have in consequence of this sole action

$$0 = m.(Py - Qx) + m'.(P'y' - Q'x')\ *;$$

thus the mutual actions of the bodies will disappear from the finite integral $\Sigma.m(Py - Qx)$. Let S be the force which attracts m towards the origin of the co-ordinates; we shall have in consequence of this sole force,

$$P = \frac{-S.x}{\sqrt{x^2 + y^2 + z^2}}\ \dagger;\quad Q = \frac{-S.y}{\sqrt{x^2 + y^2 + z^2}};$$

the force S will therefore disappear from the expression $Py - Qx$; consequently in the case in which the

---

* The notes given to the preceding number, render it unnecessary to say any thing respecting the equations given in the first part of this, as they may be proved in a similar manner. The equations $\Sigma.m(Py - Qx) = 0$, &c. are evidently true, as $\Sigma.mP = 0$, $\Sigma.mQ = 0$, and $\Sigma.mR = 0$. See the last number.

† If a radius vector is drawn from the body m to the origin of the co-ordinates, it will be equal to $\sqrt{(x^2 + y^2 + z^2)}$; by drawing a perpendicular from m to the axis of x we shall have from resolving the force S, which must be taken negatively as it tends to diminish the co-ordinates, into two others one acting in the direction of the axis and the other perpendicular to it, the following proportion,

$$\sqrt{(x^2 + y^2 + z^2)} : x :: -S : P,\text{ therefore}$$

$$P = \frac{-S.x}{\sqrt{(x^2 + y^2 + z^2)}}.$$

different bodies of the system are only solicited by their action, and their mutual attraction, and by forces directed towards the origin of the co-ordinates we shall have

$$c = \Sigma . m . \frac{xdy - ydx}{dt} \; ; \quad c' = \Sigma . m . \frac{xdz - zdx}{dt} \; ;$$

$$c'' = \Sigma . m . \frac{ydz - zdy}{dt} \quad \ldots (Z).$$

If we project the body m upon the plane of x and y, the differential $\frac{xdy - ydx}{2}$ will be the area, which the radius vector drawn from the origin of the co-ordinates to the projection of m traces during the time dt*;

---

* Suppose that A (fig. 15,) represents the origin of the co-ordinates x and y which are measured along the rectangular lines AX and AY, and N the place of a body or its projection, which is supposed to move along the curve ACN, having $AP = y$ and $PN = AX = x$ for its co-ordinates; then the area ACN described by the radius vector AP drawn from the centre A to the point N, is equal to the area $ACNP$ minus the area ANP, but the area $ACNP = \int xdy + c$, and the area $ANP = \frac{xy}{2}$, therefore the area $ACN = \int xdy + c - \frac{xy}{2}$ and consequently its differential $= xdy - d.\frac{xy}{2} = xdy - \frac{xdy}{2} - \frac{ydx}{2} = \frac{xdy - ydx}{2}$.

If $x = \varrho . \cos . \varpi$ and $y = \varrho . \sin . \varpi$, see page 21; by substitution and neglecting indefinitely small quantities of the second order, $\frac{xdy - ydx}{2} = \frac{\varrho^2 d\varpi}{2}$, which is the area of the sector described by AN during the time dt.

the sum of these areas multiplied respectively by the masses of the bodies, is therefore proportional to the element of the time; from which it follows, that in a finite time it is proportional to the time. It is this which constitutes the principle of the preservation of areas*.

The fixed plane of x and y being arbitrary, this principle has place for any plane whatever; and if the force S is nothing, that is to say, if the bodies are only subjected to their action and mutual attraction, the origin of the co-ordinates is arbitrary, and we are able to place that fixed point at will. Lastly, it is easy to perceive by what precedes, that this principle holds good in the case in which by the mutual action of the bodies of the system, sudden changes take place in their motions†.

---

* The product of the mass of a body by the area described by the projection of its radius vector during an interval of time denoted by unity, is equal to the projection of the entire force of this body multiplied by the perpendicular let fall from the fixed point upon the direction of the force thus projected. This last product is the moment of the force which would make the system turn about an axis passing through the fixed point in a perpendicular direction to the plane of projection. The principle of the preservation of areas may therefore be reduced to this; the sum of the moments of the finite forces that would make the system turn about any axis whatever, which is nothing in the case of equilibrium, is constant in that of motion. In this point of view the principle answers to all the possible laws between the force and the velocity.

It appears from what has been said, that the law of the motion of the centre of gravity and that of the areas described

There is a plane with respect to which c' and c'' are nothing which, for this reason, it is interesting to know; for it is evident that the equality of c' and c'' to nothing, ought to introduce the greatest simplicity into the research of the motion of a system of bodies. In order to determine this plane, it is necessary to refer the co-ordinates x, y, and z to three other axes, having the same origin as the preceding. Let therefore θ represent the inclination of the plane sought that is formed by two of these new axes, to the plane of x and y; and ψ the angle which the axis of x forms with the intersection of these two planes; so that $\frac{\pi}{2} - \theta$ may be the inclination of the third new axis to the plane of x and y, and $\frac{\pi}{2} - \psi$ may be the angle which its projection upon the same plane makes with the axis of x, π being the semi-circumference*.

---

being proportional to the times, (which last will again be noticed in this number) are independent of the mutual action of the bodies of the system; these two laws have therefore a more universal application than the law of the preservation of living forces, which is only independent of those passive resisting forces such as the pressures of the bodies, the tensions of the threads or rods connecting them, &c. which hinder the conditions of the system from being disturbed, or in general all those forces which can be expressed by equations between the different co-ordinates of a body, which are independent of the time.

* In *(fig. 16,)* let the lines Ax, Ay, and Az represent the axes of the three rectangular co-ordinates x, y, and z,

In order to fix our ideas, let us suppose that the origin of the co-ordinates is at the centre of the earth; that the plane of x and y is that of the ecliptic, and that the axis of z is a line drawn from the centre of the earth to the north pole of the ecliptic; let us also suppose that the plane sought is that of the equator, and that the third new axis is that of the rotation of the earth directed towards the north pole; θ will then be the obliquity of the ecliptic, and ψ the longitude of the fixed axis of x relative to the moveable equinox of spring. The two first new axes will be in the plane of the equator; and by naming φ the angular distance of the first of these axes from this equinox, φ will represent the rotation of the earth reckoned from the same equinox, and $\frac{\pi}{2} + \varphi$ will be the angular distance of the

the line Ae the common section of the planes xy and $x'y'$, and let another plane be supposed to pass through the point A perpendicular to Ae the common intersection of the planes, cutting the plane $x'y'$ in the line Ad and the plane xy in the line aAb; then as every line drawn from the point A perpendicular to the common section Ae must be in the perpendicular plane, the axes Az and Az' of the ordinates z and z' are in it, consequently as the angle dAb or θ is the inclination of the two planes and the angle dAz' is a right angle, the angle $z'Aa$ formed by the ordinate z' and the plane xy is represented by $\frac{\pi}{2} - \theta$. Also the angle xAe formed by the intersection of the planes xy and $x'y'$ and the ordinate x is represented by ψ, and the angle xAa, as the angle aAe is a right angle by $\frac{\pi}{2} - \psi$.

second of these axes from the same equinox: let these three axes be named principal axes. This being agreed upon, suppose $x_{,}$, $y_{,}$, and $z_{,}$ to be the co-ordinates of m relative, first, to the line drawn from the origin of the co-ordinates to the equinox of spring; $x_{,}$ being taken positive on the side of this equinox; secondly, to the projection of the third principal axis upon the plane of x and y; and thirdly, to the axis of z; we shall then have*

$$x = x_{,}.\cos.\psi + y_{,}.\sin.\psi;$$
$$y = y_{,}.\cos.\psi - x_{,}.\sin.\psi;$$
$$z = z_{,}.$$

---

* In (*fig.* 17,) let A be the centre of the co-ordinates, the perpendicular lines $AX_{,}$ and $AY_{,}$ the axes of the co-ordinates $x_{,}$ and $y_{,}$, and the perpendicular lines AX and AY the axes of the co-ordinates x and y making with the former lines the angle ψ, also suppose the line AM to represent the projection of the line drawn from the origin of the co-ordinates to the body m upon the plane of the x's and y's, from the point M draw the lines $Mx_{,}$ and $My_{,}$ perpendicular to the lines $AX_{,}$ and $AY_{,}$, and the lines Mx and My perpendicular to the lines AX and AY, then we shall have $Mx_{,} = y_{,}$, $My_{,} = Ax_{,} = x_{,}$, $Mx = y$ and $My = Ax = x$, also the angle $X_{,}AX = \psi$, and the angle $Aax_{,}$ formed by the intersection of the lines AX and $Mx_{,} = 90 - \psi$. In the right angled triangle $aAx_{,}$ by trigonometry $\cos.\psi$: rad.(1) : : $Ax_{,}(x_{,})$: $Aa = \dfrac{x_{,}}{\cos.\psi}$; also rad.(1) : $\sin.\psi$: : $Aa\left(\dfrac{x_{,}}{\cos.\psi}\right)$: $ax_{,}= x_{,}.\dfrac{\sin.\psi}{\cos.\psi}$; then as $Mx_{,} = y_{,}$, $Ma = y_{,} - x_{,}.\dfrac{\sin.\psi}{\cos.\psi}$. In the right angled triangle aMx, as the angle $Max = $ ang.$Aax_{,} = 90 - \psi$,

Let $x_{,,}$, $y_{,,}$, and $z_{,,}$ be co-ordinates referred, first, to the line of the equinox of spring; secondly, to the perpendicular to this line in the plane of the equator; and thirdly, to the third principal axis, then we shall have

$$x_{,} = x_{,,};$$
$$y_{,} = y_{,,}.\cos.\theta + z_{,,}.\sin.\theta;$$
$$z_{,} = z_{,,}.\cos.\theta - y_{,,}.\sin.\theta.$$

Lastly, let $x_{,,,}$, $y_{,,,}$, and $z_{,,,}$ be the co-ordinates of m referred to the first, to the second, and to the third principal axis respectively; then

$$x_{,,} = x_{,,,}.\cos.\varphi - y_{,,,}.\sin.\varphi;$$
$$y_{,,} = y_{,,,}.\cos.\varphi + x_{,,,}.\sin.\varphi;$$
$$z_{,,} = z_{,,,}.$$

From which it is easy to conclude, that

$$x = x_{,,,}.\{\cos.\theta.\sin.\psi.\sin.\varphi + \cos.\psi.\cos.\varphi\}$$
$$+ y_{,,,}.\{\cos.\theta.\sin.\psi.\cos.\varphi - \cos.\psi \sin.\varphi\} + z_{,,,}.\sin.\theta.\sin.\psi;$$

$$y = x_{,,,}.\{\cos.\theta.\cos.\psi.\sin.\varphi - \sin.\psi.\cos.\varphi\}$$
$$+ y_{,,,}.\{\cos.\theta.\cos.\psi.\cos.\varphi + \sin.\psi.\sin.\varphi\} + z_{,,,}.\sin.\theta.\cos.\psi;$$

$$z = z_{,,,}.\cos.\theta - y_{,,,}.\sin.\theta.\cos.\varphi - x_{,,,}.\sin.\theta.\sin.\varphi.$$

---

the angle $aMx = \psi$, and by trigonometry rad.(1) : sin. ψ :: $aM\left(y_{,} - x_{,}.\dfrac{\sin.\psi}{\cos.\psi}\right)$: $ax = y_{,}.\sin.\psi - x_{,}.\dfrac{\sin.^2\psi}{\cos.\psi}$ therefore $Aa + ax = x = \dfrac{x_{,}}{\cos.\psi} + y_{,}.\sin.\psi - x_{,}.\dfrac{\sin.^2\psi}{\cos.\psi} = x_{,}.\cos.\psi + y_{,}.\sin.\psi$.

Again in the triangle aMx, by trigonometry rad. (1) : cos. ψ :: $aM\left(y_{,} - x_{,}.\dfrac{\sin.\psi}{\cos.\psi}\right)$: $Mx = y = y_{,}.\cos.\psi - x_{,}.\sin.\psi$.

In a similar manner the co-ordinates relative to the other axes may be found.

By multiplying these values of x, y, and z respectively by the co-efficients of $x_{\prime\prime\prime}$ in these values, we shall have from adding them together

$x_{\prime\prime\prime} = x.\{\cos.\theta.\sin.\psi.\sin.\varphi + \cos.\psi.\cos.\varphi\}$
$+ y.\{\cos.\theta.\cos.\psi.\sin.\varphi - \sin.\psi.\cos.\varphi\} - z.\sin.\theta.\sin.\varphi.$

By multiplying in like manner the values of x, y, and z respectively by the co-efficients of $y_{\prime\prime\prime}$ in these values and afterwards by the co-efficients of $z_{\prime\prime\prime}$, we shall have

$y_{\prime\prime\prime} = x.\{\cos.\theta.\sin.\psi.\cos.\varphi - \cos.\psi.\sin.\varphi\}$
$+ y.\{\cos.\theta.\cos.\psi.\cos.\varphi + \sin.\psi.\sin.\varphi\} - z.\sin.\theta.\cos.\varphi.$
$z_{\prime\prime\prime} = x.\sin.\theta.\sin.\psi + y.\sin.\theta.\cos.\psi + z.\cos.\theta.$

These different transformations of the co-ordinates will be very useful to us as we proceed. If we place one, two, &c. marks above the co-ordinates x, y, z, $x_{\prime\prime\prime}$, $y_{\prime\prime\prime}$, and $z_{\prime\prime\prime}$, we shall have the co-ordinates corresponding to the bodies m', m'', &c.

From the above it is easy to conclude by substituting c, c', and c'' in the places of $\Sigma.m.\frac{xdy - ydx}{dt}$, $\Sigma.m.\frac{xdz - zdx}{dt}$, and $\Sigma.m.\frac{ydz - zdy}{dt}$, that

$\Sigma.m.\frac{x_{\prime\prime\prime}dy_{\prime\prime\prime} - y_{\prime\prime\prime}dx_{\prime\prime\prime}}{dt} = c.\cos.\theta - c'.\sin.\theta.\cos.\psi + c''.\sin.\theta.\sin.\psi;$

$\Sigma.m.\frac{x_{\prime\prime\prime}dz_{\prime\prime\prime} - z_{\prime\prime\prime}dx_{\prime\prime\prime}}{dt} = c.\sin.\theta.\cos.\varphi + c'.\{\sin.\psi.\sin.\varphi +$
$\cos.\theta.\cos.\psi.\cos.\varphi\}$
$+ c''.\{\cos.\psi.\sin.\varphi - \cos.\theta.\sin.\psi.\cos.\varphi\};$

$\Sigma.m.\frac{y_{\prime\prime\prime}dz_{\prime\prime\prime} - z_{\prime\prime\prime}dy_{\prime\prime\prime}}{dt} = -c.\sin.\theta.\sin.\varphi + c'.\{\sin.\psi.\cos.\varphi$
$- \cos.\theta.\cos.\psi.\sin.\varphi\}$
$+ c''.\{\cos.\psi.\cos.\varphi + \cos.\theta.\sin.\psi.\sin.\varphi\}.$

If we determine ψ and θ so that we may have $\sin.\theta.\sin.\psi = \dfrac{c''}{\sqrt{c^2+c'^2+c''^2}}$; and $\sin.\theta.\cos.\psi = \dfrac{-c'}{\sqrt{c^2+c'^2+c''^2}}$; which give

$$\cos.\theta = \dfrac{c}{\sqrt{c^2+c'^2+c''^2}}*;$$

we shall have

$$\Sigma.m.\dfrac{x_{\prime\prime\prime}dy_{\prime\prime\prime}-y_{\prime\prime\prime}dx_{\prime\prime\prime}}{dt} = \sqrt{c^2+c'^2+c''^2};$$

$$\Sigma.m.\dfrac{x_{\prime\prime\prime}dz_{\prime\prime\prime}-z_{\prime\prime\prime}dx_{\prime\prime\prime}}{dt} = 0;$$

$$\Sigma.m.\dfrac{y_{\prime\prime\prime}dz_{\prime\prime\prime}-z_{\prime\prime\prime}dy_{\prime\prime\prime}}{dt} = 0;$$

---

* Let α, β, and γ represent the angles which a perpendicular to the invariable plane respectively makes with the axes x, y, and z we shall then have the following equations,

$$\cos.\alpha = \sin.\theta.\sin.\psi,$$
$$\cos.\beta = \sin.\theta.\cos.\psi,$$
$$\cos.\gamma = \cos.\theta.$$

In order to prove that $\cos.\alpha = \sin.\theta.\sin.\psi$, let C (*fig* 18) be supposed the centre of the co-ordinates, CX the axis of x, CF the line of intersection of the planes xy and $x_{\prime\prime}y_{\prime\prime}$, CA the axis of $z_{\prime\prime}$, which is perpendicular to the line CF; from any point A of the line CA let fall the perpendicular AB upon the plane xy, join CB, then CB will be the projection of the line CA upon that plane; let the angle of the inclination of the planes xy and $x_{\prime\prime}y_{\prime\prime}$ be represented by θ, then its complement the angle ACB will be $\dfrac{\pi}{2} - \theta$, π being the semi-circumference of a circle the radius of which is unity. From A draw AD perpendicular to CX and join BD, let

the values of c' and c'' are therefore nothing with respect to the plane of x_{III} and y_{III} determined in this manner. There is only one plane which possesses this property, for supposing it is that of x and y, we shall have

$$\Sigma . m . \frac{x_{III} dz_{III} - z_{III} dx_{III}}{dt} = c . \sin.\theta . \cos.\varphi ;$$

$$\Sigma . m . \frac{y_{III} dz_{III} - z_{III} dy_{III}}{dt} = - c . \sin.\theta . \sin.\varphi .$$

---

the angle FCX be denoted by ψ, then its complement the angle XCB will be $\frac{\pi}{2} - \psi$. If AC is supposed to represent the radius equal to unity of a circle, then $BC = \sin.\theta$. and, as the triangle BDC is right angled, by trigonometry we have

rad.(1) : sin.ψ :: BC (sin.θ) : $CD = \sin.\theta.\sin.\psi$;

but CD is the cosine of the angle ACD of inclination of the axes z_{III} and x, therefore cos.$\alpha = \sin.\theta.\sin.\psi$.

From the centre C of the co-ordinates draw the line CY perpendicular to CX or the axis of x, then CY will represent the axis of y, from A let fall the perpendicular AY upon CY, join BY. In the right angled triangle BYC we have

rad.(1) : cos.ψ :: CB(sin.θ) : $CY = \sin.\theta.\cos.\psi$;

but CY is the cosine of the angle ACY or β, therefore cos.$\beta = \sin.\theta.\cos.\psi$.

The angle formed by the axes z and z_{III} is equal to that formed by the planes xy and $x_{III} y_{III}$ consequently cos.$\gamma = \cos.\theta$.

We have therefore the following equations,

$$\cos.\alpha = \sin.\theta.\sin.\psi = \frac{c''}{\sqrt{c^2 + c'^2 + c''^2}},$$

$$\cos.\beta = \sin.\theta.\cos.\psi = \frac{-c'}{\sqrt{c^2 + c'^2 + c''^2}},$$

By equalling these two functions to nothing, we shall have $\sin\theta = 0$; that is the plane of x_{III} and y_{III} then coincides with that of x and y. The value of $\Sigma.m.\frac{x_{III}dy_{III}-y_{III}dx_{III}}{dt}$ being equal to $\sqrt{c^2+c'^2+c''^2}$, whatever may be the plane of x and y; it results that the quantity $c^2+c'^2+c''^2$ is the same, whatever this plane may be, and that the plane of x_{III} and y_{III} determined by what precedes, is the plane relative to which the function $\Sigma.m.\frac{x_{III}dy_{III}-y_{III}dx_{III}}{dt}$ is the greatest*; the plane

$$\cos.\gamma = \cos.\theta = \frac{c}{\sqrt{c^2+c'^2+c''^2}},$$

from which the position of the plane may be readily found. It appears preferable to take $c' = \Sigma.m.\frac{zdx-xdz}{dt}$ instead of $\Sigma.m.\frac{xdz-zdx}{dt}$ in which case $\cos.\beta$ would be affirmative. As the quantities c, c', and c'' are constant the position of the plane is invariable.

* As the quantity $c^2+c'^2+c''^2$ is invariable whatever may be the plane of x and y, let x', y', and z' represent the co-ordinates of any other system of rectangular axes about the same point, as those of x, y, and z, then if a, d', and a'' have the same relation to the planes formed by these co-ordinates as c, c', and c'' have to those formed by the co-ordinates x, y, and z, we shall have $a^2+a'^2+a''^2 = c^2+c'^2+c''^2$, therefore $a = \sqrt{c^2+c'^2+c''^2-a'^2-a''^2}$; consequently, when a' and a'' vanish a is a maximum and equal to $\sqrt{c^2+c'^2+c''^2}$.

We may at any instant find the position of the invariable plane relative to any determinate point in space, if we know

we are treating about therefore, possesses those remarkable properties, first, that the sum of the areas traced by the projections of the radii vectores of the bodies and multiplied respectively by their masses, is the greatest possible; secondly, that the same sum relative to any plane whatever which is perpendicular to it is nothing, because the angle φ remains indeterminate. We shall be able by means of these properties to find this plane at any instant, whatever may have been the variations induced by the mutual action of the bodies in their respective positions, the same as we are able easily to find at all times the position of the centre of gravity of the system; and for this reason it is as natu-

at this instant, the velocities and the co-ordinates of all the moving bodies of the system, as the position of this plane with respect to the centre of the co-ordinates depends upon the three quantities c, c', and c'' which are respectively equal to $\Sigma m . \left(x . \frac{dy}{dt} - y . \frac{dx}{dt} \right)$, $\Sigma . m . \left(x . \frac{dz}{dt} - z . \frac{dx}{dt} \right)$ and $\Sigma . m . \left(y . \frac{dz}{dt} - z . \frac{dy}{dt} \right)$. These quantities are evidently composed of the co-ordinates of the moving bodies and the components of the velocities parallel to their axes.

It appears from the above that c is equivalent to the sum of the moments taken with reference to the centre of the co-ordinates and projected upon the plane xy of the quantities of motion of the bodies m, m', &c. at any instant; c' and c'' are the sums of the moments of these forces taken with respect to the same point and projected upon the planes of xz and yz, consequently the invariable plane coincides with the principal plane of these moments; see page 114.

ral to refer the x's and y's to this plane, as to refer the origin of the co-ordinates to the centre of gravity of the system*.

---

* If two or more bodies of the system strike each other, the position of the invariable plane will not be altered either with respect to a fixed point, if the system turns about it, or with respect to any point whatever, if the system moves freely in space. In order to prove this, it will be necessary to shew that the values of c, c', and c'' are the same after as before the impact, for which purpose the denominations given page 160 notes, will be made use of. This proof follows from the quantities of motion lost by the impact producing equilibrium, which will give the following equations,
$$\Sigma.m.\{x(b-B)-y(a-A)\}=0,$$
$$\Sigma.m.\{x(c-C)-z(a-A)\}=0,$$
$$\Sigma.m.\{y(c-C)-z(b-B)\}=0,$$
which are part of the six general equations of equilibrium No. 15, and are equivalent to the three following
$$\Sigma m(xb-ya)=\Sigma m(xB-yA),$$
$$\Sigma m(xc-za)=\Sigma m(xC-zA),$$
$$\Sigma m(yc-zb)=\Sigma m(yC-zB).$$
It is evident that the first members of these equations are the values of c, c', and c'' immediately before the impact and the second members their velocities immediately after, therefore these values are the same after as before the impact.

It therefore appears that the sums of the areas denoted by ct, $c't$, and $c''t$ are not altered by the mutual impact of the bodies of the system; these areas will consequently always be proportional to the time employed to describe them, although in the interval of that time sudden changes may have been produced in the velocities of the bodies from their mutual impact. The impact of a body not belonging to the system will in general change the values of these areas and consequently the direction of the invariable plane.

22. The principles of the preservation of living forces and of areas have place also, when the origin of

---

When a system, at liberty to turn in any direction about a fixed point, receives a number of impulses, after they have been reduced to three P, Q, and R in the directions of the axes x, y, and z, the accelerating forces $\frac{d^2x}{dt^2}$, $\frac{d^2y}{dt^2}$, and $\frac{d^2z}{dt^2}$ may be changed into the velocities $\dot{x}$, $\dot{y}$, and $\dot{z}$ and we shall have by substitution (see the beginning of this number) the following equations,

$$\left. \begin{array}{l} \Sigma.m(x\dot{y}-y\dot{x})+\Sigma.(Py-Qx)=0 \\ \Sigma.m(x\dot{z}-z\dot{x})+\Sigma.(Pz-Rx)=0 \\ \Sigma.m(y\dot{z}-z\dot{y})+\Sigma.(Qz-Ry)=0 \end{array} \right\} (a)$$

for the first instant of the motion produced by the impulses. If the system is entirely free, the point may be taken any where in space and the above equations will hold true. This will also be the case if there is no fixed point and the system turns about its centre of gravity.

If there are no accelerating forces, the effect of the impulses will be continued, the terms which depend upon the impulses P, Q, and R being regarded as constant. For, as $\dot{x}$, $\dot{y}$, and $\dot{z}$ are the velocities in directions respectively parallel to the axes of x, y, and z, we have $dx=\dot{x}dt$, $dy=\dot{y}dt$, $dz=\dot{z}dt$, &c. and the above equations will be changed into the following,

$$\Sigma.m(x\dot{y}-y\dot{x})=c,$$
$$\Sigma.m(x\dot{z}-z\dot{x})=c',$$
$$\Sigma.m(y\dot{z}-z\dot{y})=c'';$$

consequently

$$c=\Sigma.(Qx-Py),$$
$$c'=\Sigma.(Rx-Pz),$$
$$c''=\Sigma.(Ry-Qz).$$

the co-ordinates is supposed to have a rectilinear and uniform motion in space. To demonstrate it, let X, Y, and Z be named the co-ordinates of this origin, supposed in motion, referred to a fixed point, and let

$$x = X + x_i; \quad y = Y + y_i; \quad z = Z + z_i;$$
$$x' = X + x_i'; \quad y' = Y + y_i'; \quad z' = Z + z_i';$$
&c. &c. &c.

x_i, y_i, z_i, x_i', &c. will be the co-ordinates of m, m', &c. relative to the moving origin. We shall have by the hypothesis

$$d^2 X = 0; \quad d^2 Y = 0; \quad d^2 Z = 0;$$

but we have by the nature of the centre of gravity when the system is free

$$0 = \Sigma . m . \{d^2 X + d^2 x_i\} - \Sigma . m . P . dt^2;$$
$$0 = \Sigma . m . \{d^2 Y + d^2 y_i\} - \Sigma . m . Q . dt^2;$$
$$0 = \Sigma . m . \{d^2 Z + d^2 z_i\} - \Sigma . m . R . dt^2;$$

the equation (P) of No. 18, will also become by substituting $\delta X + \delta x_i$, $\delta Y + \delta y_i$, &c. instead of δx, δy, &c.;

$$0 = \Sigma . m . \delta x_i . \left\{ \frac{d^2 x_i}{dt^2} - P \right\} + \Sigma . m . \delta y_i . \left\{ \frac{d^2 y_i}{dt^2} - Q \right\} + \Sigma . m . \delta z_i . \left\{ \frac{d^2 z_i}{dt^2} - R \right\};$$

an equation which is exactly of the same form as the equation (P), if the forces P, Q, and R only depend upon the co-ordinates x_i, y_i, z_i, x_i' &c. By applying

---

The values of the constant quantities c, c', and c'' may therefore be expressed by the initial impulses given to each body, and it has been shewn that these values are the sums of the moments of these impulses with respect to the axes of x, y, and z.

to it the preceding analysis, we shall obtain the principles of the preservation of living forces and of areas with respect to the moving origin of the co-ordinates.

If the system be not acted upon by any forces unconnected with it, its centre of gravity will have a right lined and uniform motion in space, as we have seen at No. 20; by fixing therefore the origin of the co-ordinates x, y, and z at this centre, these principles will always have place, X, Y, and Z being in this case the co-ordinates of the centre of gravity, we shall have by the nature of this point,

$$0 = \Sigma.m.x_i\text{;} \quad 0 = \Sigma.m.y_i\text{;} \quad 0 = \Sigma.m.z_i\text{;}$$

which equations give

$$\Sigma.m.\frac{xdy-ydx}{dt} = \frac{XdY-YdX}{dt}.\Sigma.m + \Sigma.m.\frac{x_i dy_i - y_i dx_i}{dt}\text{;}$$

$$\Sigma.m.\frac{dx^2+dy^2+dz^2}{dt^2} = \frac{dX^2+dY^2+d^2Z}{dt^2}.\Sigma.m + \Sigma.m.\frac{dx_i^2+dy_i^2+dz_i^2}{dt^2}\text{*.}$$

---

* If $X+x_i$, $Y+y_i$, $Z+z_i$, &c. be substituted for x, y, z, &c. in the equations

$$0 = \Sigma.m.\frac{xd^2y-yd^2x}{dt^2} + \Sigma.m.(Py-Qx),$$

$$0 = \Sigma.m.\frac{xd^2z-zd^2x}{dt^2} + \Sigma.m.(Pz-Rx),$$

$$0 = \Sigma.m.\frac{yd^2z-zd^2y}{dt^2} + \Sigma.m.(Qz-Ry),$$

and regard be had to the equations

$$0 = \frac{d^2X}{dt^2}.\Sigma.m - \Sigma.mP,$$

Thus the quantities resulting from the preceding principles are composed, first, of quantities which

$$0 = \frac{d^2Y}{dt^2}.\Sigma.m - \Sigma.mQ,$$

$$0 = \frac{d^2Z}{dt^2}.\Sigma.m - \Sigma.mR,$$

we shall have the following transformations,

$$0 = \Sigma.m.\frac{x_i d^2 y_i - y_i d^2 x_i}{dt^2} + \Sigma.m.(Py_i - Qx_i),$$

$$0 = \Sigma.m.\frac{x_i d^2 z_i - z_i d^2 x_i}{dt^2} + \Sigma.m.(Pz_i - Rx_i),$$

$$0 = \Sigma.m.\frac{y_i d^2 z_i - z_i d^2 y_i}{dt^2} + \Sigma.m.(Qz_i - Ry_i),$$

which are similar to the original equations. If the quantities $Py_i - Qx_i$, &c. disappear, we shall by integration have the following equations,

$$c = \Sigma.m.\frac{x_i dy_i - y_i dx_i}{dt},$$

$$c' = \Sigma.m.\frac{x_i dz_i - z_i dx_i}{dt},$$

$$c'' = \Sigma.m.\frac{y_i dz_i - z_i dy_i}{dt}.$$

These equations are similar to those given in the last number and the same consequences may be deduced from them.

In like manner the equation

$$\Sigma.m.\left\{\left(\frac{d^2x}{dt^2}-P\right)dx + \left(\frac{d^2y}{dt^2}-Q\right)dy + \left(\frac{d^2z}{dt^2}-R\right)dz \right\} = 0,$$

may by similar substitutions be changed into the following,

$$\Sigma.m.\frac{dx_i.d^2x_i + dy_i.d^2y_i + dz_i.d^2z_i}{dt^2} - \Sigma.m(Pdx_i + Qdy_i + Rdz_i) = 0,$$

which only differs from it in having the co-ordinates referred

would have place if all the bodies of the system were united at their common centre of gravity; secondly, of quantities relative to the centre of gravity supposed immoveable; and as the first of these quantities are constant, we may see the reason why the preceding principles have place with respect to the centre of gravity. By fixing therefore at this point the origin of the co-ordinates x, y, z, x', &c. of the equations (Z) of the preceding No. they will always have place, from which it results, that the plane passing constantly through this centre and relative to which the function $\Sigma . m . \dfrac{xdy-ydx}{dt}$ is a maximum, remains always parallel to itself during the motion of the system, and that the same function relative to every other plane which is perpendicular to it is nothing.

The principles of the preservation of areas and of living forces, may be reduced to certain relations amongst the co-ordinates of the mutual distances of the bodies of the system. In fact the origin of the x's, of the y's, and of the z's being always supposed at the

---

to the centre of gravity instead of a fixed point. By integration, if the quantity $\Sigma . m . (Pdx_{\prime}+Qdy_{\prime}+Rdz_{\prime})$ be integrable and supposed equal to $d\varphi$, we shall have the following equation for the preservation of living forces with respect to the centre of gravity,

$$\Sigma . m . \frac{dx^2_{\prime}+dy^2_{\prime}+dz^2_{\prime}}{dt^2}=c+2\varphi;$$

c being a constant quantity.

centre of gravity, the equations (Z) of the preceding No. may be changed into the following forms,

$$c. \Sigma.m = \Sigma.mm'. \left\{ \frac{(x'-x).(dy'-dy)-(y'-y)}{dt} \underline{(dx'-dx)} \right\};$$

$$d. \Sigma.m = \Sigma.mm'. \left\{ \frac{(x'-x).(dz'-dz)-(z'-z)}{dt} \underline{(dx'-dx)} \right\};$$

$$c'. \Sigma.m = \Sigma.mm'. \left\{ \frac{(y'-y).(dz'-dz)-(z'-z)}{dt} \underline{(dy'-dy)} \right\};$$

It may be observed that the second members of these equations multiplied by dt, express the sum of the projections of the elementary areas traced by each right line that joins two bodies of the system, of which one is supposed to move about the other that is considered as immoveable, each area being multiplied by the product of the two masses which the right line joins.

If we apply the analysis of No. 21, to the preceding equations, we shall perceive that the plane which passes constantly through any one of the bodies of the system, and relative to which the function $\Sigma mm'. \left\{ \frac{(x'-x)(dy'-dy)-(y'-y)(dx'-dx)}{dt} \right\}$ is a maximum, remains always parallel to itself in the motion of the system; and that this plane is parallel to the plane passing through the centre of gravity relative to which the function $\Sigma.m.\frac{xdy-ydx}{dt}$ is a maximum. We shall perceive also, that the second members of the preceding equations are nothing relative to every plane which passes through the same body, and is perpendicular to the plane of which we have treated*.

---

* The positions of the invariable planes of the same system relative to two different points in space may be compared as follows; let a, β, and γ represent the co-ordinates of a new point in space, and C, C', and C'' the values of c, c', and c'' when the origin of the co-ordinates is placed at this point; we shall then have, for instance,

$$C = \Sigma . m \left\{ (x-a)\frac{dy}{dt} - (y-\beta)\frac{dx}{dt} \right\}.$$

If the sum of the masses of all the bodies be denoted by M and the co-ordinates of the centre of gravity of the system by X, Y, and Z, then

$$\Sigma . m . \frac{dy}{dt} = M . \frac{dY}{dt}, \quad \Sigma . m . \frac{dx}{dt} = M . \frac{dX}{dt};$$

therefore

$$C = \Sigma . m \left(x . \frac{dy}{dt} - y . \frac{dx}{dt} \right) - a \Sigma m . \frac{dy}{dt} + \beta \Sigma m . \frac{dx}{dt} = c + M \left(\beta . \frac{dX}{dt} - a . \frac{dY}{dt} \right).$$

In like manner

$$C' = c' + M \left(\gamma . \frac{dX}{dt} - a . \frac{dZ}{dt} \right),$$

$$C'' = c'' + M \left(\gamma . \frac{dY}{dt} - \beta . \frac{dZ}{dt} \right).$$

It is evident from the above, that if the values of a, β, and γ be such as to render the three quantities

$$\beta . \frac{dX}{dt} - a . \frac{dY}{dt}, \quad \gamma . \frac{dX}{dt} - a . \frac{dZ}{dt}, \quad \gamma . \frac{dY}{dt} - \beta . \frac{dZ}{dt},$$

respectively equal to nothing, we shall have $C = c$, $C' = c'$, and $C'' = c''$, consequently the invariable planes relative to the two centres of co-ordinates will be parallel to each other.

The equation Q of No. 19 may be put into the form
$$\Sigma.mm'.\left\{\frac{(dx'-dx)^2+(dy'-dy)^2+(dz'-dz)^2}{dt^2}\right\}=\text{const}$$
$-2\Sigma.m.\Sigma.\int mm'.Fdf;$

an equation relative solely to the co-ordinates of the mutual distances of the bodies, in which the first member expresses the sum of the squares of the relative velocities of the bodies of the system about each other, considering them two and two and supposing one of the two immoveable, each square being multiplied by the product of the two masses which we have considered.

23. By resuming the equation (R) of No. 19, and differentiating it with respect to the characteristic δ, we shall have
$$\Sigma.mv.\delta v=\Sigma.m.(P.\delta x+Q.\delta y+R.\delta z);$$

As in this case the centre of gravity of the system moves uniformly in a right line, its relative velocities $\frac{dX}{dt}$, $\frac{dY}{dt}$ and $\frac{dZ}{dt}$ are constant, therefore the three above mentioned quantities will be equal to nothing when the line joining the two centres is parallel to that described by the centre of gravity of the system. The invariable planes relative to all the points of any line parallel to that described by the centre of gravity of the system are therefore parallel to each other; and the direction of the invariable plane does not change but when it is passing from one parallel to another. It is also evident that the invariable plane relative to the centre of gravity of the system always remains parallel to itself during the motion of that centre.

the equation (P) of No. 18, then becomes

$$0 = \Sigma m . \left\{ \delta x . d . \frac{dx}{dt} + \delta y . d . \frac{dy}{dt} + \delta z . d . \frac{dz}{dt} \right\} - \Sigma . m dt . v dv,$$

Let ds be the element of the curve described by m; ds' the element of that described by m', &c.; we shall have

$$v dt = ds; \quad v' dt = ds'; \quad \&c.$$
$$ds = \sqrt{dx^2 + dy^2 + dz^2}; \quad \&c.$$

from which we may obtain by following the analysis of No. 8,

$$\Sigma . m . \delta . (v ds) = \Sigma . m . d . \frac{dx . \delta x + dy \, \delta y + dz . \delta z}{dt}.$$

By integrating this equation with respect to the differential characteristic d, and extending the integrals to the entire curves described by the bodies m, m', &c. we shall have

$$\Sigma . \delta . \int m v ds = \text{const.} + \Sigma . m . \frac{dx . \delta x + dy \, \delta y + dz \, \delta z}{dt};$$

the variations δx, δy, δz, &c. being thus but the constant quantity of the second member of this equation, relative to the extreme points of the curves described by m, m', &c.

It follows from the above, that if these points are supposed invariable, we shall have

$$0 = \Sigma . \delta . \int m v ds;$$

which shews that the function $\Sigma . \int m v ds$ is a minimum*.

---

* As $ds = v dt$, the function $\Sigma . \int m v ds$ which is a maximum or a minimum may be made to assume the form $\Sigma . m \int v^2 dt$ or $\int dt . \Sigma . m v^2$, in which $\Sigma . m v^2$ denotes the living force of the whole of the system at any instant whatever. Therefore the

It is in this that the principle of the least action in the motion of a system of bodies consists; a principle which, as we have seen, is only a mathematical result from the

principle treated upon may be properly reduced to this, that the sum of the instantaneous living forces of all the bodies from the time that they depart from certain given points until they arrive at other given points, is either a maximum or a minimum. Lagrange therefore proposes to call it the principle of the greatest or the least living force, as considered in this manner it has the advantage of being general as well for the state of motion as for that of equilibrium; for it may be proved, that the living force of a system is always the greatest or the least in the state of equilibrium, from the equation

$$\Sigma . mv^2 = c + 2\varphi, \quad (R)$$

In this equation if $\Sigma . mv^2$ is a maximum or a minimum the function φ is, generally speaking, a maximum or a minimum; consequently

$$\delta\varphi = P.\delta x + Q.\delta y + R.\delta z = 0.$$

This is the equation of the equilibrium of a system when the forces P, Q, and R are respectively functions of the lines x, y, and z; it gives the following principle, first made known by M. de Courtivron. The situation in which a system has the greatest or the least living force is that in which, if it were placed, it would remain in equilibrio.

The equilibrium would be stable if the sum of the living forces should be a maximum and unstable if a minimum; for the bodies of the system when moved from the situation of equilibrium would tend to return to it if the equilibrium were stable, their velocities would therefore diminish as they receded from it, consequently the living force would be a maximum in this position; but it would be a minimum if the equilibrium were unstable, as the bodies in receding from

primitive laws of the equilibrium and of the motion of matter. We have seen at the same time, that this principle combined with that of living forces, gives

---

their position relative to this state would tend to recede from it still farther by reason of the increase of their velocities.

In a system of bodies let g denote the force of gravity, M the sum of the masses of the bodies, and z the co-ordinate of the centre of gravity of the system, the axis of z being supposed vertical and in the direction of gravity, the equation of living forces relative to this system will be as follows,
$$\Sigma.mv^2 = c + gMz,$$
which shews that $\Sigma.mv^2$ will be a maximum or a minimum at the same time as z.

When the moving bodies are not acted upon by any accelerating forces, the sum of the living forces at each instant is constant, consequently the sum of the living forces for any time whatever is proportional to that time, from which it follows, that the system passes from one position to another in the least time possible.

In the case of perfectly hard or of perfectly elastic bodies, the sum of the products of each mass by the square of the difference between its velocities before and after the impact is a minimum. This answers to the principle of the least action.

The three equations (u), page 177, combined with that of living forces
$$\Sigma.\{m(\dot{x}^2+\dot{y}^2+\dot{z}^2)-(P\dot{x}+Q\dot{y}+R\dot{z})\}=0,$$
(see page 157.) give a property de maximis et minimis relative to a line about which the system turns in the first instant when it has received any impulse whatever, which line may be called the axis of spontaneous rotation.

the equation (P) of No. 18, which contains all that is necessary for the determination of the motions of the system.

Let a, b, and c represent the parts of the velocities $\dot{x}$, $\dot{y}$, and $\dot{z}$ which depend upon the change of position of the bodies of the system with respect to each other; when they are added to the velocities resulting from the rotations page 107, the complete values of $\dot{x}$, $\dot{y}$, and $\dot{z}$ will be expressed as follows

$$\dot{x} = z\dot{\omega} - y\dot{\varphi} + a, \quad \dot{y} = x\dot{\varphi} - z\dot{\psi} + b, \quad \dot{z} = y\dot{\psi} - x\dot{\omega} + c.$$

If these equations are differentiated, $\dot{\psi}$, $\dot{\omega}$, and $\dot{\varphi}$ being regarded as variable, we shall have

$$\delta\dot{x} = z\delta\dot{\omega} - y\delta\dot{\varphi}, \quad \delta\dot{y} = x\delta\dot{\varphi} - z\delta\dot{\psi}, \quad \delta\dot{z} = y\delta\dot{\psi} - x\delta\dot{\omega}.$$

The three equations (a) being multiplied respectively by $\delta\dot{\varphi}$, $\delta\dot{\omega}$, and $\delta\dot{\psi}$ and added together, making the variations $\delta\dot{\varphi}$, $\delta\dot{\omega}$, and $\delta\dot{\psi}$, which are the same for all the bodies, pass under the sign Σ, will give by the substitution of the preceding values

$$\Sigma . \{ m(\dot{x}\delta x + \dot{y}\delta y + \dot{z}\delta z) - (P\delta x + Q\delta y + R\delta z) \} = 0.$$

The above equation of living forces, being differentiated with respect to δ, gives

$$\Sigma . \{ 2m(\dot{x}\delta\dot{x} + \dot{y}\delta\dot{y} + \dot{z}\delta\dot{z}) - (P\delta x + Q\delta y + R\delta z) \} = 0.$$

By the comparison of these equations it is evident that

$$\Sigma . m(\dot{x}\delta\dot{x} + \dot{y}\delta\dot{y} + \dot{z}\delta\dot{z}) = 0,$$

consequently

$$\delta . \Sigma . m(\dot{x}^2 + \dot{y}^2 + \dot{z}^2) = 0.$$

This equation shews that the living force which the system acquires by impulsion, is always either a maximum or a minimum with respect to the rotations relative to three axes; and as these rotations may be composed into one about the

Lastly we have seen at No. 22, that this principle has place also, when the origin of the co-ordinates is in motion; provided that its motion be right lined and uniform and that the system be free.

---

axis of spontaneous rotation it follows, that this axis is in such a position as to have the living force of all the system the greatest or the least with respect to it.

This property of the axis of rotation with respect to solid bodies of any form was discovered by Euler, and extended by Lagrange to any system of bodies either invariably connected together or not, when these bodies receive any impulses whatever.

CHAP. VI.

Of the laws of the motion of a system of bodies in all the possible mathematical relations between the force and the velocity.

24. We have before observed at No. 5, that there are an infinite number of ways of expressing the force by the velocity, which do not imply a contradiction. The simplest of them is that of the force being proportional to the velocity, which as we have seen, is the law of nature. It is according to this law, that we have explained in the preceding chapter the differential equations of the motion of a system of bodies; but it is easy to extend the analysis of which we have made use, to all the mathematical laws possible between the velocity and the force, and thus to present under a new point of view, the general principles of motion. For this purpose, let us suppose that F being the force and v the velocity, we have $F = \varphi(v)$; $\varphi(v)$ being any function whatever of v: let us denote by $\varphi'(v)$ the differential of $\varphi(v)$ divided by dv. The denominations of the preceding numbers always remaining, the body

m will be acted upon parallel to the axis of x by the force $\varphi(v) \cdot \frac{dx}{ds}$. In the following instant this force will become $\varphi(v) \cdot \frac{dx}{ds} + d \cdot \left(\varphi(v) \cdot \frac{dx}{ds} \right)$ or $\varphi(v) \cdot \frac{dx}{ds} + d \cdot \left(\frac{\varphi(v)}{v} \cdot \frac{ds}{dt} \right)$, because $\frac{ds}{dt} = v$. Moreover P, Q, and R being the forces which act upon the body m parallel to the axes of the co-ordinates: the system will be by No. 18, in equilibrio in consequence of these forces and of the differentials $d \cdot \left(\frac{dx}{dt} \cdot \frac{\varphi(v)}{v} \right)$, $d \cdot \left(\frac{dy}{dt} \cdot \frac{\varphi(v)}{v} \right)$, $d \cdot \left(\frac{dz}{dt} \cdot \frac{\varphi(v)}{v} \right)$ taken with a contrary sign; we shall have therefore instead of the equation (P) of the same number the following;

$$0 = \Sigma . m . \left\{ \delta x . \left\{ d . \left(\frac{dx}{dt} \cdot \frac{\varphi(v)}{v} \right) - P . dt \right\} + \delta y . \right\} d . \left(\frac{dy}{dt} \cdot \frac{\varphi(v)}{v} \right) - Q . dt \right\} + \delta z . \left\{ d . \left(\frac{dz}{dt} \cdot \frac{\varphi(v)}{v} \right) - R . dt \right\} \right\} ; \quad (S)$$

which only differs in this respect, that $\frac{dx}{dt}$, $\frac{dy}{dt}$, and $\frac{dz}{dt}$ are multiplied by the function $\frac{\varphi(v)}{v}$; which may be supposed equal to unity in the case of the force being proportional to the velocity. But this difference renders the solution of the problems of mechanics very difficult. Notwithstanding we can obtain from the equation (S), certain principles analogous to those of the preservation of living forces, of areas, and of the centre of gravity.

If we change δx into dx, δy into dy, and δz into dz, &c. we shall have

$$\Sigma.m.\left\{dx.d.\left(\frac{dx}{dt}.\frac{\varphi(v)}{v}\right)+dy.d.\left(\frac{dy}{dt}.\frac{\varphi(v)}{v}\right)+dz.d.\left(\frac{dz}{dt}.\frac{\varphi(v)}{v}\right)\right\}=\Sigma.m.vdv.dt.\varphi'(v);$$

and consequently

$$\Sigma.\smallint mvdv.\varphi'(v)=\text{const.}+\Sigma.\smallint m.(P.dx+Q.dy+R.dz).$$

By supposing $\Sigma.m.(Pdx+Qdy+Rdz)$ an exact differential equal to $d\lambda$, we shall have

$$\Sigma.\smallint mvdv.\varphi'(v)=\text{const.}+\lambda; \qquad (T)$$

an equation analogous to the equation (R) of No. 19, and which changes into it in the case of nature where $\varphi'(v)=1$.

The principle of the preservation of living forces has place therefore, in all the mathematical laws possible between the force and the velocity, provided, that we understand by the living force of a body, the product of its mass by double the integral of its velocity multiplied by the differential of the function of the velocity which denotes the force.

If we suppose in the equation (S), $\delta x'=\delta x+\delta x_{\prime}'$; $\delta y'=\delta y+\delta y_{\prime}'$; $\delta z'=\delta z+\delta z_{\prime}'$; $\delta x''=\delta x+\delta x_{\prime}''$; &c.; we shall have by equalling separately to nothing the co-efficients of δx, δy, and δz,

$$0=\Sigma.m\left\{d.\left(\frac{dx}{dt}.\frac{\varphi(v)}{v}\right)-Pdt\right\}; \quad 0=\Sigma.m\left\{d.\left(\frac{dy}{dt}.\frac{\varphi(v)}{v}\right)-Qdt\right\}; \quad 0=\Sigma.m\left\{d.\left(\frac{dz}{dt}.\frac{\varphi(v)}{v}\right)-Rdt\right\}.$$

These three equations are analogous to those of No. 20, from which we have deduced the preservation of the motion of the centre of gravity in the case of nature, where the system is only subjected to the action and mutual attraction of the bodies of the system. In this case $\Sigma.mP$, $\Sigma.mQ$, and $\Sigma.mR$ are nothing and we have

$$\text{const.} = \Sigma.m.\frac{dx}{dt}.\frac{\varphi(v)}{v}; \quad \text{const.} = \Sigma.m.\frac{dy}{dt}.\frac{\varphi(v)}{v};$$

$$\text{const.} = \Sigma.m.\frac{dz}{dt}.\frac{\varphi(v)}{v};$$

$m.\frac{dx}{dt}.\frac{\varphi(v)}{v}$ is equal to $m\varphi(v).\frac{dx}{ds}$, and this last quantity is the finite force of a body resolved parallel to the axis of x; the force of a body being the product of its mass by the function of the velocity which expresses the force. Therefore the sum of the finite forces of a system resolved parallel to any axis whatever, is in this case constant whatever may be the relation of the force to the velocity; and what distinguishes the state of motion from that of rest is, that in the last state this same sum is nothing. These results are common to all the mathematical laws possible between the force and the velocity; but it is only in the law of nature that the centre of gravity moves with an uniform and rectilinear motion.

Again, let us suppose in the equation *(S)*,

$$\delta x' = \frac{y'.\delta x}{y} + \delta x_i'; \quad \delta x'' = \frac{y''.\delta x}{y} + \delta x_i''; \quad \&c.$$

$$\delta y = \frac{-x.\delta x}{y} + \delta y_i; \quad \delta y' = \frac{-x'.\delta x}{y} + \delta y_i'; \quad \&c.$$

the variation δx will disappear from the variations of the mutual distances f, f', &c. of the bodies of the system and of the forces which depend upon these quantities. If the system is free from obstacles independent of it, we shall have by equalling to nothing the co-efficient of δx,

$$0 = \Sigma.m.\left\{ x.d.\left(\frac{dy}{dt}.\frac{\varphi(v)}{v}\right) - y.d.\left(\frac{dx}{dt}.\frac{\varphi(v)}{v}\right) \right\} + \Sigma.m.\{Py - Qx\}.dt;$$

from which we may obtain by integration

$$c = \Sigma.m.\left(\frac{xdy - ydx}{dt}\right)\frac{\varphi(v)}{v} + \Sigma.\!\int\! m.(Py - Qx)dt.$$

We shall in like manner have

$$c' = \Sigma.m.\left(\frac{xdz - zdx}{dt}\right)\frac{\varphi(v)}{v} + \Sigma.\!\int\! m.(Pz - Rx).dt;$$

$$c'' = \Sigma.m.\left(\frac{ydz - zdy}{dt}\right)\frac{\varphi(v)}{v} + \Sigma.\!\int\! m.(Qz - Ry).dt;$$

c, c', and c'' being constant quantities.

If the system is only subjected to the mutual action of its parts, we have by No. 21, $\Sigma.m.(Py - Qx) = 0$; $\Sigma.m.(Pz - Rx) = 0$; $\Sigma.m.(Qz - Ry) = 0$; also, $m\left(x.\frac{dy}{dt} - y.\frac{dx}{dt}\right).\frac{\varphi(v)}{v}$ is the moment of the finite force by which the body is actuated, resolved parallel to the plane of x and y, to make the system turn about the axis of z; the finite integral $\Sigma.m.\frac{(xdy - ydx)}{dt}.\frac{\varphi(v)}{v}$ is therefore the sum of the moments of all the finite forces of the bodies of the system, which are exerted to make it turn about the same axis; this sum is therefore constant. It is nothing in the state of equilibrium; we have here therefore, the same difference between these two states, but relatively with respect to the sum of the forces parallel to any axis whatever. In the law of nature this property indicates, that the sum of the areas described about a fixed point by the projections of the radii vectores of the bodies, is always the same in equal times; but the areas described are constant, only in the law of nature.

If we differentiate, with respect to the characteristic δ, the function $\Sigma.\int m.\varphi(v).ds$; we shall have

$\delta.\Sigma.\int m.\varphi(v).ds = \Sigma.\int m.\varphi(v).\delta ds + \Sigma.\int m.\delta v.\varphi'(v).ds$;

but we have

$\delta ds = \dfrac{dx.\delta dx + dy.\delta dy + dz.\delta dz}{ds} = \dfrac{1}{v}.\left\{\dfrac{dx}{dt}.d.\delta x + \dfrac{dy}{dt}.d.\delta y + \dfrac{dz}{dt}.d.\delta z\right\}$;

we shall therefore have from integrating by parts,

$\delta.\Sigma.\int m.\varphi(v).ds = \Sigma.\dfrac{m\varphi(v)}{v}.\left\{\dfrac{dx}{dt}.\delta x + \dfrac{dy}{dt}.\delta y + \dfrac{dz}{dt}.\delta z\right\}$

$-\Sigma.\int m.\left\{\delta x.d.\left(\dfrac{dx}{dt}.\dfrac{\varphi(v)}{v}\right) + \delta y.d.\left(\dfrac{dy}{dt}.\dfrac{\varphi(v)}{v}\right) + \delta z.d.\left(\dfrac{dz}{dt}.\dfrac{\varphi(v)}{v}\right)\right\} + \Sigma.\int m.\delta v.\varphi'(v).ds.$

If the extreme points of the curves described by the bodies of the system are supposed to be fixed, the term without the sign $\int$ will disappear from this equation; we shall therefore have in consequence of the equation (S),

$\delta.\Sigma.\int m.\varphi(v) ds = \Sigma.\int m.\delta v.\varphi'(v).ds - \Sigma.\int m dt.(P\delta x + Q\delta y + R\delta z)$;

but the equation (T) differentiated with respect to δ, gives

$\Sigma.\int m.\delta v.\varphi'(v).ds = \Sigma.\int m dt.(P\delta x + Q\delta y + R\delta z)$;

we have therefore

$0 = \delta.\Sigma.\int m.\varphi(v).ds.$

This equation answers to the principle of the least action in the law of nature. $m.\varphi(v)$ is the entire force of the body m, therefore the principle comes to this, that the sum of the integrals of the finite forces of the bodies of the system, multiplied respectively by the elements of their directions is a minimum: presented in this manner it answers to all the mathematical laws possible between the force and the velocity. In the state of equilibrium, the sum of the forces multiplied by the elements of their directions is nothing in consequence of the principle of virtual velocities; what therefore distinguishes in this respect, the state of equilibrium from that of motion, is, that the same differential function which is nothing in the case of equilibrium, on being integrated gives a minimum in that of motion.

CHAP. VII.

Of the motions of a solid body of any figure whatever.

25. The differential equations of the motions of translation and rotation of a solid body, may be easily deduced from those which we have given in Chap. V; but their importance in the theory of the system of the world, induces us to develope them to a greater extent.

Let us suppose a solid body, all the parts of which are solicited by any forces whatever. Let x, y, and z be the orthogonal co-ordinates of its centre of gravity; $x + x'$, $y + y'$, and $z + z'$ the co-ordinates of any molecule dm of the body, then x', y', and z' will be the co-ordinates of this molecule with respect to the centre of gravity of the body. Let moreover P, Q, and R be the forces which solicit the molecule parallel to the axes of x, y, and z. The forces destroyed at each instant in the molecule dm parallel to these axes will be by No. 18, if the element dt of the time is supposed constant,

$$-\left(\frac{d^2x + d^2x'}{dt}\right).dm + P.dt.dm\,;$$

$$-\left(\frac{d^2y+d^2y'}{dt}\right).dm + Q.dt.dm;$$

$$-\left(\frac{d^2z+d^2z'}{dt}\right).dm + R.dt.dm.$$

It follows therefore that all the molecules acted upon by similar forces should mutually cause an equilibrium. We have seen at No. 15, that for this purpose it is necessary that the sum of the forces parallel to the same axis should be nothing, which gives the three following equations;

$$S.\left(\frac{d^2x+d^2x'}{dt^2}\right).dm = S.Pdm;$$

$$S.\left(\frac{d^2y+d^2y'}{dt^2}\right).dm = S.Qdm;$$

$$S.\left(\frac{d^2z+d^2z'}{dt^2}\right).dm = S.Rdm;$$

the letter S being here a sign of integration relative to the molecule dm, which ought to be extended to the whole mass of the body. The variables x, y, and z are the same for all the molecules, we may therefore suppose them independent of the sign S; thus denoting the mass of the body by m, we shall have

$$S.\frac{d^2x}{dt^2}.dm = m.\frac{d^2x}{dt^2}; \quad S.\frac{d^2y}{dt^2}.dm = m.\frac{d^2y}{dt^2};$$

$$S.\frac{d^2z}{dt^2}.dm = m.\frac{d^2z}{dt^2}.$$

We have moreover by the nature of the centre of gravity

$$S.x'.dm = 0; \quad S.y'.dm = 0; \quad S.z'.dm = 0;$$

which equations give

$$S.\frac{d^2x'}{dt^2}.dm = 0; \quad S.\frac{d^2y'}{dt^2}.dm = 0; \quad S.\frac{d^2z'}{dt^2}.dm = 0;$$

we shall therefore obtain

$$m.\frac{d^2x}{dt^2} = S.Pdm\,;$$
$$m.\frac{d^2y}{dt^2} = S.Qdm\,;\quad\quad (A)$$
$$m.\frac{d^2z}{dt^2} = S.Rdm\,;$$

these three equations determine the motion of the centre of gravity of a body, and answer to the equations of No. 20, relative to the centre of gravity of a system of bodies.*

We have seen at No. 15, that for the equilibrium of a solid body, the sum of the forces parallel to the axis of x multiplied respectively by their distances from the axis of z, minus the sum of the forces parallel to the axis of y multiplied by their distances from the axis of z, is equal to nothing; we shall therefore have

---

* As the equations (A) do not contain the co-ordinates x', x'', &c. of the different molecules of the body, they are independent of them and only indicate the motion of the centre of gravity of the body. This motion is not influenced by the mutual actions of the molecules upon each other, but solely by the accelerating forces which solicit them.

It is evident from the above that the centre of gravity of any free body whatever, like that of a system, has always the same motion as if this body were all concentrated into one point and acted upon by the same accelerating forces as the parts of the body were, when in their natural state. This accords with what has been given at No. 20.

$$S. \left\{ (x+x').\frac{d^2y+d^2y'}{dt^2} - (y+y').\frac{d^2x+d^2x'}{dt^2} \right\}.dm =$$
$$S.\{(x+x').Q - (y+y').P\}.dm ; \qquad (1)$$
but we have
$$S.(x.d^2y - y.d^2x).dm = m.(xd^2y - yd^2x) ;$$
and in like manner
$$S.(Qx - Py).dm = x.S.Qdm - y.S.Pdm ;$$
lastly we have
$$S.\{x'd^2y + xd^2y' - y'd^2x - yd^2x'\}.dm = d^2y.S.x'dm - d^2x.S.y'dm + x.S.d^2y'.dm - y.S.d^2x'.dm ;$$
and by the nature of the centre of gravity each of the terms of the second member of this equation is nothing; the equation (1) will therefore become in consequence of the equations (*A*),
$$S.\left(\frac{x'd^2y' - y'd^2x'}{dt^2}\right).dm = S.(Qx' - Py').dm ;$$
by integrating this equation with respect to the time *t*, we shall have
$$S.\frac{x'dy' - y'dx'}{dt}.dm = S.\int(Qx' - Py').dt.dm ;$$
the sign $\int$ of integration being relative to the time *t*.

From the above it is easy to conclude, that if we make
$$S.\int(Qx' - Py').dt.dm = N ;$$
$$S.\int(Rx' - Pz').dt.dm = N' ;$$
$$S.\int(Ry' - Qz').dt.dm = N'' ;$$
we shall have the three following equations
$$\left. \begin{array}{l} S.\dfrac{x'dy' - y'dx'}{dt}.dm = N ; \\[4pt] S.\dfrac{x'dz' - z'dx'}{dt}.dm = N' ; \\[4pt] S.\dfrac{y'dz' - z'dy'}{dt}.dm = N'' ; \end{array} \right\} ; \qquad (B)$$

these three equations contain the principle of the preservation of areas; they are sufficient for determining the motion of the rotation of a body about its centre of gravity*; united to the equations (A) they completely determine the motions of the translation and of the rotation of a body.

If the body be forced to turn about a fixed point; it results from No. 15, that the equations (B) are sufficient for this purpose; but it is then necessary to fix the origin of the co-ordinates x', y', and z' at this point.

26. Let us particularly consider these equations and suppose the origin fixed at any point whatever, different or not from the centre of gravity. Let us refer the position of each molecule to three axes perpendicular to each other and fixed in the body, but moveable in space. Let θ be the inclination of the plane formed by the two first axes upon the plane of x' and y'; let φ be the angle formed by the line of intersection of these two planes and the first axis; lastly, let ψ be the com-

---

* As the equations (B) do not contain the co-ordinates of the centre of gravity of the system, they only shew the different positions of the body with respect to three axes which have their origin at that centre, consequently the motion of rotation which the equations (B) determine, is the same as if the centre of gravity were at rest.

A body acted upon by accelerating forces may therefore have two motions; one that of translation, the same as if the body were concentrated into one point at its centre of gravity and acted upon by all the forces parallel to their directions, and the other that of rotation about the centre of gravity the same as if this point were fixed.

plement of the angle which the projection of the third axis upon the plane of x and y makes with the axis of x. We will name these three new axes principal axes, and we shall denote by x'', y'', and z'' the three co-ordinates of the molecule dm referred to these axes; then, by No. 21, the following equations will have place,

$x' = x''.\{\cos.\theta.\sin.\psi.\sin.\varphi + \cos.\psi.\cos.\varphi\}$
$\quad + y''.\{\cos.\theta.\sin.\psi.\cos.\varphi - \cos.\psi.\sin.\varphi\} + z''.\sin.\theta.\sin.\psi$;

$y' = x''.\{\cos.\theta.\cos.\psi.\sin.\varphi - \sin.\psi.\cos.\varphi\}$
$\quad + y''.\{\cos.\theta.\cos.\psi.\cos.\varphi + \sin.\psi.\sin.\varphi\} + z''.\sin.\theta.\cos.\psi$;

$z' = z''.\cos.\theta - y''.\sin.\theta.\cos.\varphi - x''.\sin.\theta.\sin.\varphi$.

By means of these equations we shall be enabled to develope the first members of the equations *(B)* in functions of θ, ψ, and φ, and of their differentials. But we may considerably simplify the calculations, by observing that the position of the three principal axes depends upon three constant quantities, which can always be determined so as to satisfy the three equations

$S.x''y''.dm = 0$; $\quad S.x''z''.dm = 0$; $\quad S.y''z''.dm = 0$.

Suppose then

$S.(y''^2 + z''^2).dm = A$; $\quad S.(x''^2 + z''^2).dm = B$;
$\quad\quad S.(x''^2 + y''^2).dm = C$;

and for abridgment let

$d\varphi - d\psi.\cos.\theta = p dt$;
$d\psi.\sin.\theta.\sin.\varphi - d\theta.\cos.\varphi = q dt$;
$d\psi.\sin.\theta.\cos.\varphi + d\theta.\sin.\varphi = r dt$.

The equations *(B)* will, after all the reductions, be changed into the three following,

$$\left.\begin{array}{l}A.q.\sin.\theta.\sin.\varphi+Br.\sin.\theta.\cos.\varphi-Cp.\cos.\theta=\\-N;\\ \cos.\psi.\{Aq.\cos.\theta.\sin.\varphi+Br.\cos.\theta.\cos.\varphi+\\ Cp.\sin.\theta\}\\ +\sin.\psi.\{Br.\sin.\varphi-Aq.\cos.\varphi\}=-N';\\ \cos.\psi.\{Br.\sin.\varphi-Aq.\cos.\varphi\}\\ -\sin.\psi.\{Aq.\cos.\theta.\sin.\varphi+Br.\cos.\theta.\cos.\varphi\\ +Cp.\sin.\theta\}=-N'';\end{array}\right\}; (C)$$

these three equations give by differentiating them and supposing $\psi=0$ after the differentiations, which is equivalent to taking the axis of the x's indefinitely near to the line of intersection of the plane of x' and y' with that of x'' and y'',

$d\theta.\cos.\theta.(Br.\cos.\varphi+Aq.\sin.\varphi)+\sin.\theta.d.(Br.\cos.\varphi+Aq.\sin.\varphi)-d.(Cp.\cos.\theta)=-dN;$

$d\psi.(Br.\sin.\varphi-Aq.\cos.\varphi)-d\theta.\sin.\theta.(Br.\cos.\varphi+Aq.\sin.\varphi)+\cos.\theta.d.(Br.\cos.\varphi+Aq.\sin.\varphi)+d.(Cp.\sin.\theta)$
$=-dN';$

$d.(Br.\sin.\varphi-Aq.\cos.\varphi.)-d\psi.\cos.\theta.(Br.\cos.\varphi+Aq.\sin.\varphi)-Cpd\psi.\sin.\theta=-dN'';$

If we make
$$Cp=p'; \quad Aq=q'; \quad Br=r';$$
these three differential equations will give the following

$$\left.\begin{array}{l}dp'+\dfrac{B-A}{AB}.q'r'.dt=dN.\cos.\theta-dN'.\sin.\theta;\\[4pt] dq'+\dfrac{C-B}{CB}.r'p'.dt=-(dN.\sin.\theta+dN'.\cos.\\ \qquad\qquad\qquad\qquad\theta).\sin.\varphi+dN''.\cos.\varphi;\\[4pt] dr'+\dfrac{A-C}{CA}.p'q'.dt=-(dN.\sin.\theta+dN'\cos.\\ \qquad\qquad\qquad\qquad\theta).\cos.\varphi-dN''.\sin.\varphi.\end{array}\right\}; (D)$$

these equations are very convenient for determining the motion of rotation of a body when it turns very nearly about one of its principal axes, which is the case of the celestial bodies.

27. The three principal axes to which we have referred the angles θ, ψ, and φ, deserve particular attention. Let us proceed to determine their position in any solid whatever.

The values of x', y', and of z' the preceding number, give by No. 21, the following equations,

$x'' = x'.(\cos.\theta.\sin.\psi.\sin.\varphi + \cos.\psi.\cos.\varphi)$
$\qquad + y'.(\cos.\theta.\cos.\psi.\sin.\varphi - \sin.\psi.\cos.\varphi) - z'.\sin.\theta.\sin.\varphi;$

$y'' = x'.(\cos.\theta.\sin.\psi.\cos.\varphi - \cos.\psi.\sin.\varphi)$
$\qquad + y'.(\cos.\theta.\cos.\psi.\cos.\varphi + \sin.\psi.\sin.\varphi) - z'.\sin.\theta.\cos.\varphi;$

$z'' = x'.\sin.\theta.\sin.\psi + y'.\sin.\theta.\cos.\psi + z'.\cos.\theta.$

From which may be obtained

$x''\cos.\varphi - y''.\sin.\varphi = x'.\cos.\psi - y'.\sin.\psi;$
$x''\sin.\varphi + y''.\cos.\varphi = x'.\cos.\theta.\sin.\psi + y'.\cos.\theta.\cos.\psi - z'.\sin.\theta.$

Suppose

$S.x'^2.dm = a^2;\quad S.y'^2.dm = b^2;\quad S.z'^2.dm = c^2;$
$S.x'y'.dm = f;\quad S.x'z'.dm = g;\quad S.y'z'.dm = h;$

then

$\cos.\varphi.S.x''z''.dm - \sin.\varphi.S.y''z''.dm = (a^2 - b^2).\sin.\theta.\sin.\psi.\cos.\psi + f.\sin.\theta.(\cos.^2\psi - \sin.^2\psi) + \cos.\theta.(g.\cos.\psi - h.\sin.\psi);$

$\sin.\varphi.S.x''z''.dm + \cos.\varphi.S.y''z''.dm = \sin.\theta.\cos.\theta.(a^2.\sin.^2\psi + b^2.\cos.^2\psi - c^2 + 2f.\sin.\psi.\cos.\psi)$
$\qquad + (\cos.^2\theta - \sin.^2\theta).(g.\sin.\psi + h.\cos.\psi).$

By equalling to nothing the second members of these two equations, we shall have

$$\tan \theta = \frac{h.\sin\psi - g.\cos\psi}{(a^2-b^2).\sin\psi.\cos\psi + f.(\cos^2\psi - \sin^2\psi)};$$

$$\tfrac{1}{2}\tan 2\theta = \frac{g.\sin\psi + h.\cos\psi}{c^2 - a^2.\sin^2\psi - b^2.\cos^2\psi - 2f.\sin\psi.\cos\psi};$$

but we have

$$\tfrac{1}{2}\tan 2\theta = \frac{\tan\theta}{1-\tan^2\theta};$$

by equalling these values of $\tfrac{1}{2}\tan 2\theta$, and substituting in the last instead of $\tan\theta$ its preceding value in ψ, and then making for abridgment $\tan\psi = u$, we shall obtain, after all the reductions, the following equation of the third degree;

$$0 = (gu+h).(hu-g)^2$$
$$+\{(a^2-b^2).u+f.(1-u^2)\}.\{(hc^2-ha^2+fg).u+gb^2-gc^2-hf\}.$$

As this equation has at least one real root, it is evidently always possible to render equal to nothing at the same time, the two quantities

$$\cos\varphi.S.x''z''.dm - \sin\varphi.S.y''z''.dm;$$
$$\sin\varphi.S.x''z''.dm + \cos\varphi.S.y''z''.dm;$$

and consequently the sum of their squares, $(S.x''z''.dm)^2 + (S.y''z''.dm)^2$, which requires that we should have separately

$$S.x''z''.dm = 0; \quad S.y''z''.dm = 0.$$

The value of u gives that of the angle ψ, and consequently that of the $\tan\theta$, and of the angle θ. It is yet required to determine the angle φ, which may be done by means of the condition $S.x''y''.dm = 0$, which remains to be fulfilled. For this purpose it may be observed, that if we substitute in $S.x''y''.dm$ for x'' and y'' their preceding values, that function will be changed into the form, $H.\sin 2\varphi + L.\cos 2\varphi$, H and L being functions of the angles θ and ψ, and of the constant quantities a^2, b^2, c^2, f, g, h,; by equalling this expression to nothing, we shall have

$$\tan.2\varphi = \frac{-L}{H}.\ *$$

The three axes determined by means of the preceding values of θ, ψ, and φ, satisfy the three equations

$$S.x''y''.dm = 0\ ;\ S.x''z''.dm = 0\ ;\ S.y''z''.dm = 0\ \dagger;$$

---

* If F be supposed equal to $x'\cos.\psi - y'\sin.\psi$ and G to $x'\cos.\theta.\sin.\psi + y'\cos.\theta.\cos.\psi - z'\sin.\theta$, then $x'' = F.\cos.\varphi + G.\sin.\varphi$ and $y'' = G.\cos.\varphi - F.\sin.\varphi$; consequently

$$x''y'' = (G^2 - F^2)\sin.\varphi.\cos.\varphi + FG.(\cos.^2\varphi - \sin.^2\varphi),$$

therefore

$$S.x''y''.dm = \sin.\varphi.\cos.\varphi.S(G^2 - F^2)dm + (\cos.^2\varphi - \sin.^2\varphi)S.FG.dm.$$

If the second member of this equation be equalled to nothing, as $2\sin.\varphi.\cos.\varphi = \sin.2\varphi$ and $\cos.^2\varphi - \sin.^2\varphi = \cos.2\varphi$, we shall have

$$\sin.2\varphi.S.(G^2 - F^2)\cdot dm + 2\cos.2\varphi.S.FG.dm = 0.$$

If $H = S.(G^2 - F^2).dm$, $L = 2.S.FG.dm$ and $\frac{\sin.2\varphi}{\cos.2\varphi} = \tan.2\varphi$, by substitution, the above equation may be changed into the following

$$\tan.2\varphi = \frac{-L}{H}.$$

† These calculations which would be found very tedious in practice are much facilitated by the knowledge of one of the principal axes of rotation. Thus for instance, let the position of the axis x'' be known in the body, as the situation of the three rectangular axes x', y', and z' are arbitrary, x'' may be supposed to coincide with x' which will cause the angles φ and ψ and consequently their sines to vanish, their

The equation of the third degree in u seems to indicate three systems of principal axes similar to the preceding; but it ought to be observed, that u is the tangent of the angle formed by the axis of x' and the intersection of

co-sines will then be equal to unity, also the quantities f and g will be equal to nothing, as is evident from substituting in them for y' and z' their respective values. The equation
$$\tfrac{1}{2}\tan 2\theta = \frac{g\sin\psi + h\cos\psi}{c^2 - a^2\sin^2\psi - b^2\cos^2\psi - 2f\sin\psi\cos\psi}$$
may therefore be transformed into the following $\tan 2\theta' = \dfrac{2h'}{c'^2 - b'^2}$, in which h', c', and b' are the respective values of h, c, and b when x'' coincides with x'. It appears from this expression, as $\tan 2\theta' = \tan 2(\theta' + 90)$, that the other two axes must be taken in the plane of $y'\,z'$, one making the angle θ' and the other the angle $\theta' + 90$ with the axis of y' or the plane of $x''\,y'$.

If $h' = 0$ and $b' = c'$ the angle θ becomes indeterminate, therefore every line in the plane of $y'\,z'$ passing through the origin of the axes is a perpendicular axis.

When the bodies are symmetrically formed, the axis of the figure is always a principal axis and the others may be found by this rule. Thus for instance in the case of the ellipsoid which has three unequal conjugate diameters, let them be taken for the axes of x, y, and z, then the body will be divided into similar and equal parts by each of the planes of the co-ordinates; therefore each molecule dm above the plane of xy having the co-ordinates x, y, and z will have another corresponding and equal one below that plane having x, y, and $-z$ for its co-ordinates; consequently the indefinitely small elements of the integrals $S.xz.dm$ and $S.yz.dm$ corresponding to these molecules will be $xz.dm$ and $-xz.dm$, $yz.dm$ and $-yz.dm$, therefore the integrals will vanish as

the plane of x' and y' with that of x'' and y''; but it is evident that it is possible to change into each other the three axes of x'' of y'' and of z'', as the three preceding equations will be always satisfied; the equation in u ought therefore equally to determine the tangent of the angle formed by the axis of x' and the intersection of the plane of x' and y', either with the plane of x'' and y'', or with the plane of x'' and z'', or with the plane of y'' and z''. Thus the three roots of the equation in u are real and appertain to the same system of axes.

It follows from the above that, generally, a solid has only one system of axes which possess the property treated upon. These axes have been named the principal axes of rotation, by reason of a property that is peculiar to them, which will be noticed in the course of this work*.

---

they may be supposed to consist of an indefinitely great number of indefinitely small quantities, which from their contrary signs destroy each other. The integral $S.xy.dm$ may in a similar manner be proved equal to nothing, consequently the three axes of the ellipsoid are principal axes.

* These axes are called the principal axes of rotation on account of the property which the body possesses, if not acted upon by any accelerating force, of always turning round any one of them, if once put in motion about it in consequence of an initial impulse. This property may be demonstrated in the following manner.

Suppose a body, not acted upon by any accelerating force, to turn about a fixed axis in consequence of an impulse which has been given it. Let one of the three rectangular coordinates to which the molecules of the body are referred,

The sum of the products of each molecule of a body into the square of its distance from an axis, is called its moment of inertia with respect to this axis. Thus the

---

that of z for instance be supposed to pass along this axis, conceive r to represent the distance of a molecule dm from it, having its angular velocity, which is common to all the points of the body, denoted by ω. The centrifugal force of the molecule dm will be represented by $r\omega^2$, consequently the moving force with which the molecule acts upon the fixed axis in a perpendicular direction to its length is $r\omega^2.dm$. The resultant or the two resultants of the whole number of forces which act upon the axis, shew the pressure which it sustains.

To find this pressure let us suppose that the force $r\omega^2.dm$ is applied directly to the axis of z where its direction meets it. The force may be resolved at the point where it acts upon the axis into two others parallel to the axes of x and y, the cosines of the angles that the direction of the force $r\omega^2.dm$ makes with the axes x and y are $\dfrac{x}{r}$ and $\dfrac{y}{r}$, consequently the components of this force parallel to these axes are $x\omega^2.dm$ and $y\omega^2.dm$, therefore the resultant or sum of all the components parallel to the axis of x is the integral $S.x\omega^2.dm$ or $\omega^2 S.x.dm$; this quantity, if M represent the mass of the body and X the value of x at its centre of gravity, is equal to $\omega^2 MX$. In like manner the resultant of the forces parallel to the axis of y directed in the plane of yz is $\omega^2 MY$, Y representing the value of y at the centre of gravity of the body. Let z' and z'' denote the distances of these resultants from the plane of x and y, then by the theory of moments we shall have the following equations

$$MXz' = S.xz.dm, \quad MYz'' = S.yz.dm.$$

which will give the values of z' and z''.

quantities A, B, and C are the moments of inertia of the solid which we have considered, with respect to

---

If z' and z'' are equal, the forces ωMX and ωMY are applied at the same point, and may consequently be reduced to one $\omega^2.M\sqrt{X^2+Y^2}$, which represents the action upon the fixed axis arising from the centrifugal force of the body. Let the axis of z be one of three principal axes which have their origin at the centre of gravity of the body at the distance a from the origin of z upon that axis; in this case $X=0$ and $Y=0$; let the origin of the co-ordinates be moved without changing their directions from its first point to the centre of gravity of the body, then the co-ordinates of dm will be x, y, and $z-a$. As z is one of the principal axes, we shall have the following equations.
$$S.x(z-a).dm = S.xz.dm - aS.xdm = 0,$$
$$S.y(z-a).dm = S.yz.dm - aS.ydm = 0,$$
therefore because $S.xdm = MX = 0$ and $S.ydm = MY = 0$, it is evident that $S.xz.dm = 0$ and $S.yz.dm = 0$, consequently $\omega^2 S.xz.dm = 0$ and $\omega^2 S.yz.dm = 0$. As the resultant $\omega^2.MX$ of forces directed in the plane of xz, and the sum $\omega^2 S.xz.dm$ of their moments are nothing, these forces will be in equilibrio independent of the fixed axis. The same is evidently the case with respect to the forces in the plane of y and z; therefore in this case the centrifugal forces of the different molecules of the body do not act upon the axis of rotation and the same motion would be continued about it, if it were not fixed.

If the fixed axis z is a principal one of rotation that does not pass through the centre of gravity of the body X and Y will not vanish, but as $S.xz.dm$ and $S.yz.dm$ are equal to nothing, the distances z' and z'' must consequently be equal to nothing, therefore the fixed axis will be acted upon by the force $\omega^2.M\sqrt{X^2+Y^2}$ applied at the origin of the co-ordinates; if this point be fixed the pressure arising from

the axes x'', y'', and z''.* Let us name C' the moment of inertia of the same solid with respect to the

---

the centrifugal force will be destroyed, and if the axis of rotation were free to move about this point the rotatory motion with respect to it would still be continued.

It is evident therefore, that if any point in a body be taken as a fixed one; there will be three axes passing through it, round which the body would move uniformly without acting upon them.

If the body were forced to turn about any other axis passing through that point, the action of the centrifugal forces upon it would not take place at that point and it would consequently be displaced. Vide the Mechanique Philosophique of Prony, a similar proof is also given by Poisson.

* Suppose for example that MA (*fig. 2*) is a parallelepiped it is required to find its moment of inertia with respect to the axis MB. Let $MG=a$, $ME=b$, and $MB=c$ and suppose the axes of the co-ordinates x, y, and z to be taken from their origin at M in the respective directions of these lines, then if the uniform density of the parallelepiped is represented by ϱ and $dx\,dy\,dz$ is the volume of a molecule of the body, its inertia will be denoted by the integral $SSS.(x^2+y^2)\varrho dxdydz$. This quantity when integrated with respect to z, from $z=0$ to $z=c$, gives
$$\varrho c SS.(x^2+y^2).dxdy.$$
This expression when integrated with respect to y, from $y=0$ to $y=b$, becomes
$$\varrho c S.\left(x^2 b + \frac{b^3}{3}\right)dx;$$
from which by integrating with respect to x, from $x=0$ to $x=a$ the result
$$\varrho c\left(\frac{a^3 b}{3} + \frac{ab^3}{3}\right)$$

axis of z', and we shall find by means of the values of x' and of y' of the preceding number

---

may be obtained, which is the moment of inertia of the parallelepiped MA with respect to the axis MB. The moments with respect to the other axes may be obtained by properly changing amongst themselves the letters a, b, and c; let $M = \varrho.abc$, then the moments with respect to the three axes MG, ME, and MB will be

$$\frac{M}{3}(b^2+c^2), \quad \frac{M}{3}(a^2+c^2), \quad \frac{M}{3}(a^2+b^2).$$

The moments of inertia of an homogeneous ellipsoid with respect to the three principal diameters, may be readily found from the general equation to its surface

$$a^2b^2z^2 + a^2c^2y^2 + b^2c^2x^2 = a^2b^2c^2,$$

a, b, and c representing the lengths of the three rectangular semi-diameters, which are respectively in the directions of the three co-ordinates x, y, and z. The inertia with respect to the axis of z is

$$\varrho SSS(x^2+y^2)dxdydz,$$

ϱ representing the density of the body.

This expression when integrated with respect to the entire ellipsoid gives

$$\frac{4}{15}.abc\varrho\pi.(a^2+b^2)$$

π denoting the ratio of the circumference to the diameter of a circle. By changing the letters a and c into each other the moment of inertia

$$\frac{4}{15}.abc\varrho\pi.(b^2+c^2)$$

will be obtained with respect to the axis of x. In like manner by changing a and b into each other the moment of inertia

$$\frac{4}{15}.abc\varrho\pi(a^2+c^2)$$

$C = A.\sin.^2\theta.\sin.^2\varphi + B.\sin.^2\theta.\cos.^2\varphi + C.\cos.^2\theta.$

The quantities $\sin.^2\theta.\sin.^2\varphi$, $\sin.^2\theta.\cos.^2\varphi$, and $\cos.^2\theta$ are the squares of the cosines of the angles which the axes of x'', y'', and z'' make with the axis of z'; from which it follows in general, that if we multiply the moment of inertia relative to each principal axis of rotation, by the square of the cosine of the angle that it makes with any axis whatever, the sum of these three products will be the moment of inertia of the solid relative to this last axis.

The quantity C' is less than the greatest and greater will be obtained with respect to the axis of y. The content of the ellipsoid is represented by $\frac{4\pi}{3}.abc$, and its mass M by $\varsigma.\frac{4\pi}{3}.abc$: by substitution therefore the following will be the moments of inertia of the ellipsoid with respect to the axes x, y, and z,

$$\frac{M}{5}(b^2+c^2), \quad \frac{M}{5}(a^2+c^2), \quad \frac{M}{5}(a^2+b^2).$$

The greatest moment of inertia is about the shortest and the least about the longest axis.

In the case of the spheroid $b = c$ and the moments with respect to the axes of x and y are $\frac{2M}{5}.b^2$ and $\frac{M}{5}(a^2+b^2)$.

In the case of the sphere $a = b = c$, and the moments of inertia with respect to any axis passing through its centre is $\frac{8\pi}{15}.\varsigma a.^5$.

than the least of the three quantities A, B, and C*; the greatest and the least moments of inertia therefore appertain to the principal axes†.

---

* Suppose A to be the greatest and C the least of the quantities A, B, and C, and α, β, and γ to denote the three angles which the axis z' respectively makes with the axes x'', y'', and z'', then by observing that $\cos.^2\alpha + \cos.^2\beta \cos.^2\gamma = 1$, notes No. 2, the preceding equation may be made to assume the following forms,

$$C' = A - (A-B)\cos.^2\beta - (A-C)\cos.^2\gamma,$$
$$C' = C + (A-C)\cos.^2\alpha + (B-C)\cos.^2\beta;$$

which shew that C' is less than the greatest and greater than the least of the three quantities A, B, and C.

† A method of finding one of the principal axes of rotation of a body may be derived from the properties which two of the axes possess, of having the moments of inertia with respect to one of them a maximum and to the other a minimum. Thus let x'' be an axis of rotation passing through the origin of the co-ordinates which it is required to find, and suppose it makes the angle θ with the plane $x'y'$, and that its projection upon the plane $x'y'$ makes the complement of the angle ψ with the axis of x'. Suppose also the axis of y'' to be drawn perpendicular to x'' in the plane of $x'y'$, and the axis of z'' to be perpendicular to the plane $x''y''$, all the co-ordinates having the same origin, then the following values of x'', y'', and z'' may be obtained,

$$x'' = z'\sin.\theta + (y'\cos.\psi + x'\sin.\psi)\cos.\theta,$$
$$y'' = y'\sin.\psi - x'\cos.\psi,$$
$$z'' = z'\cos.\theta - (y\cos.\psi + x\sin.\psi)\sin.\theta.$$

Let these values of y'' and z'' be substituted in the expression $S(y''^2 + z''^2)dm$ which may be denoted by L, then if the angles θ and ψ are supposed variable we shall find

Let X, Y, and Z be the co-ordinates of the centre of gravity of the solid referred to the origin of the co-ordinates, which we shall fix at the point about which the body is forced to turn, if it be not free; $x'-X$, $y'-Y$ and $z'-Z$ will be the co-ordinates of the molecule dm of the body relative to its centre of gravity: the moment of inertia relative to an axis parallel to the axis of z' and passing through the centre of gravity, will therefore be

$$S.\{(x'-X)^2+(y'-Y)^2\}.dm$$

but we have by the nature of the centre of gravity, $S.x'dm = mX$; $S.y'dm = mY$; the preceding moment will therefore be reduced to

$$-m.(X^2+Y^2)+S.(x'^2+y'^2).dm.$$

We shall consequently have the moments of inertia of a solid relative to the axes which pass by any point whatever, when these moments shall be known relative to the axes which pass through the centre of gravity. It is evident at the same time, that the least of all the

---

$$\left(\frac{dL}{d\theta}\right) = -2S.(x''z'').dm$$

$$\left(\frac{dL}{d\psi}\right) = -2\cos.\theta.S(x''y'').dm.$$

If these values be equalled to nothing they will give L either a maximum or a minimum: the angles θ and ψ may then be obtained from them which will shew the position of the axis x'', from which the other two may be readily found. This method like that of Laplace requires very tedious calculations.

moments of inertia has place with respect to one of the three principal axes which pass through this centre*.

---

* The moment of inertia with respect to any axis z' is $S.(x'^2+y'^2).dm$, but for any other axis parallel to z' which has the lines a and b for the co-ordinates of any one of its points, this expression becomes

$$S.(x'^2+y'^2).dm - 2a.S.x'.dm - 2b.S.y'dm + (a^2+b^2).m.$$

Let r be equal to the distance between the axes or $\sqrt{a^2+b^2}$ then if the axis of z' passes through the centre of gravity of the body, as $S.x'dm=0$ and $S.y'dm=0$, No. 15, the expression will become

$$S.(x'^2+y'^2).dm + r^2 m;$$

therefore the moment of inertia with respect to an axis passing through the centre of gravity of a body, is less than for any other axis, by the square of the distance of the two axes multiplied into the mass of the body, consequently the minimum minimorum of the moments of inertia of a body belongs to one of the principal axes which passes through its centre of gravity.

To find those points of a body, if there be any, about which all the moments of inertia are equal.

Let a, b, and c denote the co-ordinates of one of these points, the centre of gravity of the body being taken for the origin of the co-ordinates which are supposed to be in the directions of the principal axes passing through it, then $x-a$, $y-b$, and $z-c$ will be the co-ordinates of any molecule with respect to the point sought; now from the nature of the question every straight line passing through this point must be a principal axis, consequently we have the following equations

$S.(x-a)(y-b).dm = S.xy.dm - aS.y.dm - b.S.x.dm + ab.S.dm = 0,$

If it be supposed that by the nature of the body, the two moments of inertia A and B are equal, we shall have
$$C' = A.\sin.^2\theta + C.\cos.^2\theta\ ;$$

$S.(x-a)(z-c).dm = S.xz.dm - aS.z.dm - c.S.x.dm + ac.S.dm = 0,$
$S.(y-b)(z-c).dm = S.yz.dm - b.S.z.dm - c.S.y.dm + bc.S.dm = 0\ ;$
but $S.xy.dm$, $S.xz.dm$, $S.yz.dm$, $S.x.dm$, $S.y.dm$, and $S.z.dm$ are respectively equal to nothing, therefore the above are reduced to these
$$abm = 0,\quad acm = 0,\quad bcm = 0.$$

It is evident, that if the point sought exist, as from the last equations two of the quantities a, b, and c are equal to nothing, it must be upon one of the principal axes belonging to the centre of gravity of the body. Suppose $b = c = 0$, then a the distance of the point sought from the centre of gravity is indeterminate and upon the axis of x. The moment of inertia of this point with respect to the axis of x is A, but with respect to axes parallel to those of y and z it is $B + ma^2$ and $C + ma^2$. The problem requires that we should have
$$B + ma^2 = C + ma^2 = A.$$
These equations are impossible unless $B = C$, which gives
$$a^2 = \frac{A-C}{m},$$
therefore
$$a = \pm\sqrt{\frac{A-C}{m}}\ ;$$
consequently a has two values which, if A be greater than C, are real and upon the axis of x at equal distances on each side of the centre of gravity.

It appears from the above that there cannot be any point in a body about which all the moments of inertia are equal,

by making θ equal to a right angle, which will cause the axis of z' to be perpendicular to that of z'', the equation will give $C = A$. The moments of inertia relative to all the axes situated in the plane perpedicular to the axis of z'', will be therefore equal to each other. But it is easy to be assured that, in this case, we shall have for the system of the axis of z'' and of any two axes perpendicular to it and to each other

$$S.x'y'.dm = 0; \quad S.x'z''.dm = 0; \quad S.y'z''.dm = 0;$$

for, from denoting by x'' and y'' the co-ordinates of a molecule dm of a body referred to the two principal axes taken in the plane perpendicular to the axis of z'', with respect to which the moments of inertia are supposed equal, we shall have

$$S.(x''^2 + z''^2).dm = S.(y''^2 + z''^2).dm;$$

---

if the quantities A, B, and C belonging to the centre of gravity of the body are unequal; if one of the three quantities A, B, and C is greater than either of the others, and the others equal, in this case, there are two points with respect to which all the moments of inertia are equal upon the principal axis to which the greatest of the moments A, B, and C belongs. If the three moments A, B, and C are are equal, the centre of gravity of the body is the only point about which all the moments of inertia are equal.

For example, in the oblate spheroid the points are upon the minor axis of the generating ellipse, at the distance of the square root of the fifth part of the difference between the squares of the semi-major and semi-minor axes on each side from the centre of the spheroid. This last problem was first solved by M. Binet, and afterwards in a manner similar to the above by S. D. Poisson.

or simply $S.x'^2.dm = S.y'^2.dm$; but by naming ε the angle which the axis of x' makes with the axis of x'', we have

$$x' = x''.\cos.\varepsilon + y''.\sin.\varepsilon;$$
$$y' = y''.\cos.\varepsilon - x''.\sin.\varepsilon;$$

consequently

$S.x'y'.dm = S.x''y''.dm.(\cos.^2\varepsilon - \sin.^2\varepsilon) + S.(y''^2 - x''^2).dm.\sin.\varepsilon.\cos.\varepsilon = 0.$

We shall find in like manner $S.x'z'.dm = 0$; $S.y'z''dm = 0$; all the axes perpendicular to that of z'' are therefore principal axes, and in this case the solid has an infinite number of principal axes.

If at the same time $A = B = C$, we shall have generally $C = A$; that is to say, all the moments of inertia of the solid are equal; but then we have generally

$S.x'y'.dm = 0$; $S.x'z'.dm = 0$; $S.y'z'.dm = 0$;

whatever may be the position of the plane of x' and y', so that all the axes are principal axes. This is the case of the sphere: we shall find in the course of the Méchanique Céleste that this property belongs to an infinite number of other solids of which the general equation will be given.

28. The quantities p, q, and r which we have introduced into the equations (C) of No. 26, have this remarkable property, that they determine the position of the real and instantaneous axis of rotation of a body with respect to the principal axes. In fact, we have relative to the points situated in the axis of rotation $dx' = 0$, $dy' = 0$, and $dz' = 0$; by differentiating the values of x', y', and z' of No. 26, and making the sine $\psi = 0$ after the differentiation, which may be done, because we are able to fix at will the position of the axis of x' upon the plane of x' and y', we shall have

$dx' = x''.\{d\psi.\cos.\theta.\sin.\varphi - d\varphi.\sin.\varphi\} + y''.\{d\psi.\cos.\theta.\cos.\varphi - d\varphi.\cos.\varphi\} + z''.d\psi.\sin.\theta = 0;$

$dy' = x''.\{d\varphi.\cos.\theta.\cos.\varphi - d\theta.\sin.\theta.\sin.\varphi - d\psi.\cos.\varphi\} + y''.\{d\psi.\sin.\varphi - d\varphi.\cos.\theta.\sin.\varphi - d\theta.\sin.\theta.\cos.\varphi\} + z''.d\theta.\cos.\theta = 0;$

$dz' = -x''.\{d\theta.\cos.\theta.\sin.\varphi + d\varphi.\sin.\theta.\cos.\varphi\} - y''.\{d\theta.\cos.\theta.\cos.\varphi - d\varphi.\sin.\theta.\sin.\varphi\} - z''.d\theta.\sin.\theta = 0.$

If we multiply the first of these equations by $-\sin.\varphi$, the second by $\cos.\theta.\cos.\varphi$, and the third by $-\sin.\theta.\cos.\varphi$; we shall have from adding them together,

$$0 = px'' - qz''.$$

If we multiply the first of the same equations by $\cos.\varphi$, the second by $\cos.\theta.\sin.\varphi$, and the third by $-\sin.\theta.\sin.\varphi$; we shall have from adding them together,

$$0 = py'' - rz''.$$

Lastly, if we multiply the second of the same equations by $\sin.\theta$, and the third by $\cos.\theta$, we shall have from adding them together,

$$0 = qy'' - rx''.$$

This last equation evidently results from the two preceding ones; thus the three equations $dx' = 0$, $dy' = 0$, and $dz' = 0$ are reduced to these two equations which belong to a right line forming with the axes of x'', y'', and z'', angles which have for their cosines

$$\frac{q}{\sqrt{p^2+q^2+r^2}}, \quad \frac{r}{\sqrt{p^2+q^2+r^2}}, \quad \frac{p}{\sqrt{p^2+q^2+r^2}}.\;*$$

---

* Let the right line be represented by $\sqrt{x''^2+y''^2+z''^2}$, then by letting fall a perpendicular from the end of it upon the axis of x'', we shall have by trigonometry $\sqrt{x''^2+y''^2+z''^2}$: x'' :: rad. (1) : $\dfrac{x''}{\sqrt{x''^2+y''^2+z''^2}}$ or the cosine of the angle

This right line is therefore at rest and forms the real axis of rotation of the body.

In order to have the velocity of rotation of the body, let us consider that point in the axis of z'' which is at a distance equal to unity from the origin of the co-ordinates. We shall have its velocities parallel to the axes of x', y', and z', by making $x''=0$, $y''=0$, and $z''=1$ in the preceding expressions of dx', dy', and dz', and dividing them by dt; which gives for these partial velocities

$$\frac{d\psi}{dt}.\sin.\theta\ ;\quad \frac{d\theta}{dt}.\cos.\theta\ ;\quad \frac{-d\theta}{dt}.\sin.\theta\ ;$$

the whole velocity of the point is therefore $\dfrac{\sqrt{d\theta^2+d\psi^2.\sin.^2\theta}}{dt}$ or $\sqrt{q^2+r^2}$. If we divide this velocity by the distance of the point from the instantaneous axis of rotation, we shall have the angular velocity of rotation of the body; but this distance is evidently equal to the sine of the angle which the real axis of rotation makes with the axis of z'', the cosine of which angle is which the line makes with the axis of x''; by substituting for y'' and z'' their respective values $\dfrac{rx''}{q}$ and $\dfrac{px''}{q}$, this expression becomes $\sqrt{\dfrac{x''}{x''^2+\dfrac{r^2x''^2}{q^2}+\dfrac{p^2x''^2}{q^2}}}$ or $\sqrt{\dfrac{q}{p^2+q^2+r^2}}$.

The cosines of the other angles may be found in a similar manner.

$\frac{p}{\sqrt{p^2+q^2+r^2}}$; we shall therefore have $\sqrt{p^2+q^2+r^2}$ for the angular velocity of rotation*.

It appears from the above, that whatever may be the movement of rotation of a body, either about a fixed point, or one considered as such; this movement can only be regarded as one of rotation about an axis fixed during one instant, but which may vary from one instant to another.

The position of this axis with respect to the three principal axes and the angular velocity of rotation, depend upon the variables p, q, and r; the determin-

---

* To find the angular velocity about the immoveable axis of rotation; from the distance equal to unity upon the axis of z'' let fall a perpendicular upon the axis of rotation; the perpendicular will represent the sine of the angle which this axis makes with z'', and is consequently equal to

$$\sqrt{1-\frac{p^2}{p^2+q^2+r^2}} \text{ or } \sqrt{\left(\frac{q^2+r^2}{p^2+q^2+r^2}\right)}:$$ the angular velocity about the axis of rotation at a distance represented by unity, may therefore be found by the following proportion

$$\sqrt{\left(\frac{q^2+r^2}{p^2+q^2+r^2}\right)} : 1 :: \sqrt{q^2+r^2} : \sqrt{p^2+q^2+r^2}.$$

If the quantities p, q, and r are constant the axis of rotation will remain fixed in the body, the angular velocity will also be invariable; but the converse of this is not equally true, for the axis of rotation may change its position in the body and the angular velocity remain the same, that is, the quantity $\sqrt{p^2+q^2+r^2}$ may continue constant although p, q, and r vary.

ation of which is very important in these researches, and which by expressing quantities independent of the situation of the plane of x' and y', are themselves independent of this situation.

29. Let us determine these variables in functions of the time, in the case in which the body is not solicited by any exterior forces. For this purpose let the equations (D) be resumed of No. 26, containing the variables p', q', and r' which are in a constant ratio to the preceding. The differentials dN, dN', and dN'' are in this case nothing, and these equations give by being added together after they have been respectively multiplied by p', q', and r'

$$0 = p'dp' + q'dq' + r'dr',$$

which becomes from integration

$$p'^2 + q'^2 + r'^2 = k^2;$$

k being a constant quantity.

If the equations (D) are multiplied respectively $AB.p'$, $BC.q'$, and $AC.r'$, and afterwards added together, they will give by integrating their sum

$$AB.p'^2 + BC.q'^2 + AC.r'^2 = H^2;$$

H being a constant quantity; this equation contains the principle of the preservation of living forces. From these two integrals the following equations may be obtained,

$$q'^2 = \frac{AC.k^2 - H^2 + A.(B-C).p'^2}{C.(A-B)};$$

$$r'^2 = \frac{H^2 - BC.k^2 - B.(A-C).p'^2}{C.(A-B)};$$

thus q' and r' will be known in functions of the time t, when p' shall have been determined: but the first of the equations (D) gives

$$dt = \frac{AB.dp'}{(A-B).q'r'};$$

which, by substituting the values of q' and r', becomes

$$dt = \frac{A.B.C.dp'}{\sqrt{\{AC.k^2 - H^2 + A.(B-C).p'^2\}.\{H^2 - BC.k^2 - B.(A-C).p'^2\}}};$$

an equation that is only integrable in one of the three following cases, $B = A$, $B = C$, or $A = C$.[*]

---

[*] In the cases in which this equation can be integrated it may be made to assume the following forms, in which a and b are substituted for the constant quantities.

First. If $A = B$, $dt = b . \dfrac{a^2 dp'}{a^2 + p'^2}$ and $t = b$. (circular arc having a for radius and p' for tangent) + const.

Second. If $A = C$, $dt = b . \dfrac{dp'}{\sqrt{a^2 + p'^2}}$ and $t = b$. hyp. log. $(p' + \sqrt{a^2 + p'^2})$ + const.

Third. If $B = C$, $dt = b, \dfrac{a dp'}{\sqrt{a^2 - p'^2}}$ and $t = b$. (circular arc having a for radius and p' for sine) + const.

Fourth. If $AC.k^2 = H^2$, $dt = b \dfrac{2a dp'}{p'\sqrt{a^2 - p'^2}}$ and $t = b$. hyp. log. $\dfrac{a - \sqrt{a^2 - p'^2}}{a + \sqrt{a^2 - p'^2}}$ + const.

Fifth. If $H^2 = BC.k^2$, $dt = b . \dfrac{2a dp'}{p'\sqrt{a^2 + p'^2}}$ and $t = b$. hyp. log. $\dfrac{\sqrt{a^2 + p'^2} - a}{\sqrt{a^2 + p'^2} + a}$ + const.

The determination of the three quantities p', q', and r', requires three constant quantities, H^2, k^2, and that which is introduced by the integration of the preceding equation. But these quantities only give the position of the instantaneous axis of rotation of the body upon its surface, or relative to the three principal axes and its angular velocity of rotation. To have the real movement of the body about a fixed point, it is necessary also to know the position of the principal axes in space; this should introduce three new constant quantities relative to the primitive position of these axes, and requires three new integrals, which when joined to the preceding will give the complete solution of the problem. The equations (C) of No. 26 contain the three constant quantities N, N', and N''; but they are not entirely distinct from the constant quantities H and k. In fact, if we add together the squares of the first members of the equations (C), we shall have

$$p'^2 + q'^2 + r'^2 = N^2 + N'^2 + N''^2.$$

which gives $k^2 = N^2 + N'^2 + N''^2$.

The constant quantities N, N', and N'', answer to the constant quantities c, c', and c'' of No. 21, and the function $\frac{1}{2}t\sqrt{p'^2 + q'^2 + r'^2}$ expresses the sum of the areas described during the time t by the projections of each molecule of the body upon the plane relative to which this sum is a maximum. N' and N'' are nothing relative to this plane; by therefore equalling to nothing their values found in No. 26, we shall have

$0 = Br.\sin.\varphi - Aq.\cos.\varphi$;

$0 = Aq.\cos.\theta.\sin.\varphi + Br.\cos.\theta.\cos.\varphi + Cp.\sin.\theta$;

from which may be obtained

$$\cos.\theta = \frac{p'}{\sqrt{p'^2 + q'^2 + r'^2}};$$

$$\sin.\theta.\sin.\varphi = \frac{-q'}{\sqrt{p'^2+q'^2+r'^2}};$$

$$\sin.\theta.\cos.\varphi = \frac{-r'}{\sqrt{p'^2+q'^2+r'^2}}*.$$

By means of these equations, we shall know the values of θ and φ' in functions of the time relative to the fixed

---

* If from the centre of the co-ordinates a perpendicular be erected to the invariable plane, and α, β, and γ denote the respective angles which it makes with the co-ordinates x'', y'', and z'', then it may be readily proved, No. 21 notes, that $\cos.\alpha = -\sin.\theta.\sin.\varphi$, $\cos.\beta = -\sin.\theta.\cos.\varphi$, and $\cos.\gamma = \cos.\theta$, consequently we have the following equations

$$\cos.\alpha = \frac{q'}{\sqrt{p'^2+q'^2+r'^2}},$$

$$\cos.\beta = \frac{r'}{\sqrt{p'^2+q'^2+r'^2}},$$

$$\cos.\gamma = \frac{p'}{\sqrt{p'^2+q'^2+r'^2}}.$$

The position of the invariable plane with respect to three axes fixed in space may be found in the following manner. Let O (*fig.* 19) represent the centre of the co-ordinates or the point about which the body turns, Ox'', Oy'', and Oz'' the three principal axes to which the co-ordinates x'', y'', and z'' are referred, Om the perpendicular to the invariable plane, and Ox one of the three rectangular fixed axes belonging to x, y, and z to which the ordinates x are referred. From any point x in the axis Ox let the right line xm be drawn cutting the line Om at m, then in the triangle xmO

$$xm^2 = Ox^2 + Om^2 - 2Ox.Om.\cos.xOm.$$

The co-ordinates of the point x with respect to the axes of

plane that we have considered. It only remains to find the angle ψ, which the intersection of this plane and that of the two principal axes makes with the axis of x', which requires a new integration.

The values of q and r of No. 26 give
$$d\psi.\sin.^2\theta = qdt.\sin.\theta.\sin.\varphi + rdt.\sin.\theta.\cos.\varphi\,;$$
from which may be obtained
$$d\psi = \frac{-K.dt.(Bq'^2 + Ar'^2)}{AB.(q'^2 + r'^2)}\,;$$

---

x'', y'', and z'' are $xO.\cos.xOx''$, $xO.\cos.xOy''$, and $xO.\cos.xOz''$ and those of m to the same co-ordinates are $mO.\cos.mOx''$, $mO.\cos.mOy''$, and $mO.\cos.mOz''$, we therefore have, page 8,
$xm^2 = (xO.\cos.xOx'' - mO.\cos.mOx'')^2 + (xO.\cos.xOy'' - mO.\cos.mOy'')^2 + (xO.\cos.xOz'' - mO.\cos.mOz'')^2$.
From these two values of xm^2 we shall find, by making the co-efficients of Ox^2, Om^2 and $Ox.Om$ in the equations equal to each other, that those of $Ox.Om$ give
$\cos.mOx = \cos.xOx''.\cos.mOx'' + \cos.xOy''.\cos.mOy'' + \cos.xOz''.\cos.mOz''$. In a similar manner it may be proved that $\cos.mOy = \cos.yOx''.\cos.mOx'' + \cos.yOy''.\cos.mOy'' + \cos.yOz''.\cos.mOz'$, and $\cos.mOz = \cos.zOx''.\cos.mOx'' + \cos.zOy''.\cos.mOy'' + \cos.zOz''.\cos.mOz''$. (See page 111). We therefore evidently have the following equations

$$\cos.mOx = \frac{p'.\cos.xOz'' + q'.\cos.xOx'' + r'.\cos.xOy''}{\sqrt{p'^2 + q'^2 + r'^2}},$$

$$\cos.mOy = \frac{p'.\cos.yOz'' + q'.\cos.yOx'' + r'\cos.yOy''}{\sqrt{p'^2 + q'^2 + r'^2}},$$

$$\cos.mOz = \frac{p'.\cos.zOz'' + q'.\cos.zOx'' + r.\cos.zOy''}{\sqrt{p'^2 + q'^2 + r'^2}},$$

but by what precedes

$$q'^2 + r'^2 = k^2 - p'^2; \quad Bq'^2 + Ar'^2 = \frac{H^2 - AB.p'^2}{C};$$

we shall therefore have

$$d\psi = \frac{-k.dt.\{H^2 - AB.p'^2\}}{ABC.(k^2 - p'^2)}.$$

If we substitute instead of dt its value found above; we shall have the value of ψ in a function of p'; the three angles θ, φ, and ψ will thus be determined in functions of the variables p', q', and r', which are themselves determined in functions of the time t. We shall therefore know at any instant whatever the values of these angles with respect to the plane of x' and y', which we have considered; and it will be easy by the formulæ of spherical trigonometry to find the values of the same angles relative to any other plane; this will introduce two new constant quantities, which united to the four preceding ones will form the six constant quantities, that ought to give the complete solution of the problem about which we have treated. But it is evident that the consideration of the plane above mentioned simplifies this problem.

The position of the three principal axes upon the surface of the body being supposed to be known; if at any instant whatever we are acquainted with the position of the real axis of rotation upon this surface, and the angular velocity of rotation; we shall have at this instant the values of p, q, and r, because these values divided by the angular velocity of rotation express the cosines of the angles which the real axis of rotation forms with the three principal axes: we shall therefore have the values of p', q', and r'; but these last values are proportional to the sines of the angles which the three principal axes form with the plane of x' and y', relative to which the sum of the areas of the projections of the molecules of the body multiplied respectively by these molecules, is a maximum; we shall therefore be able to determine at all times the intersection of the surface of the body by this invariable plane, and consequently to find the position of this plane by the actual conditions of the movement of the body.

Let us suppose that the movement of rotation of a body is owing to an initial impulse, which does not pass through its centre of gravity. It results from what has been demonstrated in numbers 20 and 22, that the centre of gravity will take the same motion as if this impulse was immediately applied to it, and that the body will take the same movement of rotation about this centre as if it were immoveable. The sum of the areas described about this point by the radius vector of each molecule projected upon a fixed plane and multiplied respectively by these molecules, will be proportional to the moment of the initial force projected upon the same plane, but this moment is the greatest relative to the plane which passes by its direction and by the centre of gravity; that plane is therefore the invariable one. If the distance of the initial impulse from the centre of gravity is denoted by f, and the velocity which it impresses upon this point by v; m representing the mass of the body, mfv will be the moment of this impulse, which being multiplied by $\frac{1}{2}t$ will give a product equal to the sum of the areas described during the time t; but this sum by what precedes is $\frac{1}{2}t.\sqrt{p'^2+q'^2+r'^2}$; we have therefore

$$\sqrt{p'^2+q'^2+r'^2}=m.fv.$$

If we had known at the commencement of the movement, the position of the principal axes relative to the invariable plane, or the angles θ and φ; we should also have known at this commencement, the values of p', q', and r', and consequently those of p, q, and r; we shall therefore have at any instant whatever the values of these same quantities*.

---

* The diagram *(fig. 20)* may serve to assist the learner in the readier understanding of this number.

Let the body be supposed to be put in motion about the point O situated upon the principal axis z'', by an impulse acting upon the point B of the surface in the direction of the line AB. Let MBM' be a section of the body made by a plane passing by the point O and the right line AB: this plane is the invariable one; let Om be a perpendicular to it. If we know the section MBM' of the body at the beginning of the motion, we shall know the angles which its perpendicular makes with the three principal axes. Let Ox'', Oy'', and Oz'' represent the three principal axes and OI the instantaneous axis of rotation of the body, then at the beginning of the motion

$$\cos. IOx'' = \frac{q}{\sqrt{p^2+q^2+r^2}},$$

$$\cos. IOy'' = \frac{r}{\sqrt{p^2+q^2+r^2}},$$

$$\cos. IOz'' = \frac{p}{\sqrt{p^2+q^2+r^2}}.$$

Let MOM' be the section of the planes MBM' and $x''Oy''$; then one of the constant quantities which belong to the values of t and ψ in functions of p will depend upon the time when t commenced, and the other upon the line taken arbitrarily in the plane MBM' from which the angle ψ commenced.

LAPLACE'S MECHANICS. 231

This theory may serve to explain the two motions of the rotation and revolution of the planets, by one sole initial impulse. Let us suppose that a planet is an homogeneous sphere having a radius R, and that it turns about the sun with an angular velocity U;

In the case of the rotatory movement of a solid body not acted upon by any accelerating forces we evidently have, (see page 107,)

$$dx = zd\omega - yd\varphi,$$
$$dy = xd\varphi - zd\psi,$$
$$dz = yd\psi - xd\omega.$$

If the three equations (Z) No. 21, are respectively multiplied by $\frac{d\varphi}{dt}$, $\frac{d\omega}{dt}$, and $\frac{d\psi}{dt}$, which are made to pass under the sign Σ, and added together, when $\frac{dx}{dt}$, $\frac{dy}{dt}$, and $\frac{dz}{dt}$ are substituted for their values, they will give

$$Sm \cdot \frac{dx^2 + dy^2 + dz^2}{dt^2} = c'' \cdot \frac{d\psi}{dt} + c' \cdot \frac{d\omega}{dt} + c \cdot \frac{d\varphi}{dt},$$

but as the equation (Q) No. 19, when $\varphi = 0$, gives

$$Sm \cdot \frac{dx^2 + dy^2 + dz^2}{2dt^2} = C,$$

we shall have

$$c'' \cdot \frac{d\psi}{dt} + c' \cdot \frac{d\omega}{dt} + c \cdot \frac{d\varphi}{dt} = 2C.$$

In this equation c, c', and c'' represent the initial forces of impulsion, and C a constant quantity which is necessarily positive.

If $a.\cos.\alpha$, $a.\cos.\beta$, and $a.\cos.\gamma$ are respectively substituted for c'', c', and c, (see notes No. 21) and $\frac{d\theta}{dt} \cdot \cos.\lambda$,

r being imagined to denote its distance from the sun; we shall then have $v = rU$; moreover if we suppose that the planet moves in consequence of an initial impulse, the direction of which took place at a distance f,

---

$\frac{d\theta}{dt}.\cos.\mu$, and $\frac{d\theta}{dt}.\cos.\nu$ for $\frac{d\psi}{dt}$, $\frac{d\omega}{dt}$, and $\frac{d\varphi}{dt}$ (see notes page 108) the above equation will be changed into the following,

$$\frac{d\theta}{dt} \cdot \Big(\cos.\alpha.\cos.\lambda + \cos.\beta.\cos\mu + \cos.\gamma.\cos.\nu \Big) = \frac{2C}{a}.$$

In this equation α, β, and γ are the angles which the perpendicular axis to the invariable plane makes with the fixed axes of x, y, and z; and λ, μ, and ν are the angles which the instantaneous axis of the composed rotation makes with the same axes, $\frac{d\theta}{dt}$ being the velocity of rotation. Let σ represent the angle which the instantaneous axis of rotation makes with the perpendicular axis to the invariable plane, then, (notes page 227)

$\cos.\sigma = \cos.\alpha.\cos.\lambda + \cos.\beta.\cos.\mu + \cos.\gamma.\cos.\nu$;

consequently $\frac{d\theta}{dt}.\cos.\sigma = \frac{2C}{a}$, $\frac{2C}{a}$ being a constant quantity which depends upon the initial movement of the body. We therefore have a ratio, independent of the form of the body, between the real velocity of rotation at each instant and the position of the axis of rotation relative to the invariable plane. This curious property was discovered by Lagrange.

If the plane of $x y$ be taken by the centre of the body and the right line in the direction of which the impulse is given, the constant quantities c' and c'' will vanish, and the general equation found above will be reduced to $c.\frac{d\varphi}{dt} = 2C$; which shews that the velocity of rotation with respect to the axis of z, that is parallel to the plane of the impulse, is invariable.

from its centre, it is evident that it will revolve about an axis perpendicular to the invariable plane; by therefore considering this axis as the third principal axis, we shall have $\theta = 0$, and consequently $q' = 0$, $r' = 0$; we shall therefore have $p' = mfv$ or $Cp = mfrU$.

But is well known that in the sphere we have $C = \frac{2}{5} mR^2$, consequently

$$f = \frac{2}{5} \cdot \frac{R^2}{r} \cdot \frac{p}{U};$$

which gives the distance f of the direction of the initial impulse from the centre of the planet, and answers to the relation observed between the angular velocity p of rotation, and the angular velocity U of revolution about the sun. Relative to the earth we have $\frac{p}{U} = 366,25638$; the parallax of the sun gives $\frac{R}{r} = 0,000042665$ and consequently $f = \frac{1}{100} \cdot R$, very nearly.

As the planets are not homogeneous, they may be considered as formed of spherical and concentrical laminæ of unequal densities. Let ϱ represent the density of one of these laminæ of which the radius is R, ϱ being a function of R; we shall then have

$$C = \frac{2m}{3} \cdot \frac{\int \varrho \cdot R^4 \cdot dR}{\int \varrho \cdot R^2 \cdot dR},*$$

---

* In order to find the value of C in the case where the density ϱ of each spherical lamina varies as some function of its radius, let us suppose (*fig.* 21) that $ACBD$ is the section

m being the entire mass of the planet, and the integrals taken from $R=0$, to its value at the surface; we shall therefore have

$$f = \frac{2}{3} \cdot \frac{p}{rU} \cdot \frac{\int \varrho . R^4 . dR}{\int \varrho . R^2 . dR}.$$

If, as it is natural to suppose, the laminæ be densest nearest the centre, $\dfrac{\int \varrho . R^4 . dR}{\int \varrho . R^2 . dR}$ will be less than $\frac{3}{5}R^2$; the value of f will therefore be less than in the case of homogeneity.

---

of a sphere made by a plane passing through its centre O and axis of rotation CD, let the radius OA be drawn perpendicular to CD or axis of z, and RS parallel to it or cutting the circumference of the circle $ABCD$ in R and S; draw the radius OR meeting the line SR at R, suppose $OR=R$, $OP\sqrt{x^2+y^2}=u$, then $PR=\sqrt{R^2-u^2}$, and if $p=3.14159$, &c. the surface of the cylinder generated by the revolution of SR about CD is $4pu\sqrt{R^2-u^2}$, therefore $4pudu\sqrt{R^2-u^2}$ is the differential of the solid generated by the revolution of the plane $CRSD$ about CD, consequently $4p\int u^3 du\sqrt{R^2-u^2}$ is the integral of that solid, when it has each of its molecules multiplied into the square of its distance from the axis CD; this integral may be readily found by supposing $R^2-u^2=w^2$ or $u^2=R^2-w^2$, then $u^4=R^4-2R^2w^2+w^4$, consequently $u^3 du = -R^2 w dw + w^3 dw$ and $4pu^3du\sqrt{R^2-u^2} = 4p.(-R^2 w^2 dw + w^4 dw)$, which by integration becomes $4p.(-\frac{1}{3}R^2w^3 +\frac{1}{5}w^5)+$Cor.; when $u=0$ this integral should vanish but as $x=R$ in this case $4p.(-\frac{1}{3}R^5+\dfrac{R^5}{5}+$ Cor. $=0$, there-

30. Let us now determine the oscillations of a body in the case in which it turns very nearly about the third principal axis. It is possible to deduce them from the integral in the preceding number, but it is simpler to obtain them directly from the differential equations (*D*) of No. 26. The body not being solicited by any forces, these equations will become by substituting in the places of p', q', and r' their values Cp, Aq, and Br,

$$dp + \frac{B-A}{C}.qr.dt = 0;$$

$$dq + \frac{C-B}{A}.rp.dt = 0;$$

$$dr + \frac{A-C}{B}.pq.dt = 0.$$

The solid being supposed to turn very nearly about its

---

fore, as when $u = R$, $w = 0$ the above integral for the sphere whose radius $= R$ becomes $\frac{8}{15}pR^5$. If R is supposed to be variable in this last expression, its differential will be $\frac{8}{3}pR^4 dR$, which is the value of an indefinitely small lamina of a sphere at the distance R from the centre, which has each of its molecules multiplied into the square of its distance from an axis passing through that centre; if $\rho = f.(R)$ represent the density of the lamina, then $\frac{8}{3}p\rho R^4 dR$ will denote the number of its molecules each multiplied into the square of its distance from the axis, therefore $C = \frac{8}{3}p\int\rho R^4 dR$. Now the mass m of the sphere is equal to $4p\int\rho R^2 dR$, therefore the value of p is $\frac{m}{4}.\frac{1}{\int\rho R^2 dR}$, which by substitution gives

$$C = \frac{2m}{3}.\frac{\int\rho R^4.dR}{\int\rho R^2.dR}.$$

third principal axis, q and r are very small quantities*, the squares and the products of which may be neglected; this gives $dp=0$, and consequently p a constant quantity. If in the two other equations we suppose

$$q = M.\sin.(nt+\gamma)\,;\quad r=M'.\cos.(nt+\gamma)\dagger\,;$$

we shall have

$$n=p.\sqrt{\frac{(C-A)(C-B)}{AB}}\,;\quad M'=-M.\sqrt{\frac{A.(C-A)}{B.(C-B)}}\,;$$

M and γ being two constant quantities. The angular velocity of rotation will be $\sqrt{p^2+q^2+r^2}$, or simply p, by neglecting the squares of q and r; this velocity will

---

* If the angle IOz'' (*fig.* 20) is very small the angles IOx'' and IOy'' will be very nearly right angles, therefore their cosines $\dfrac{q}{\sqrt{p^2+q^2+r^2}}$ and $\dfrac{r}{\sqrt{p^2+q^2+r^2}}$, and consequently the quantities q and r will be very small.

† By substituting $M.\sin.(nt+\gamma)$ for q and $M'.\cos.(nt+\gamma)$ for r in these equations, they will be changed into the following;

$$M.\cos.(nt+\gamma)ndt + \frac{C-B}{A}.M'.\cos.(nt+\gamma).pdt=0,$$

$$-M'.\sin.(nt+\gamma)ndt + \frac{A-C}{B}.M.\sin.(nt+\gamma).pdt=0.$$

Consequently

$$Mn + M'.\frac{C-B}{A}.p=0,$$

$$-M'n + M.\frac{A-C}{B}.p=0;$$

from which $n=p\sqrt{\dfrac{(C-A)(C-B)}{AB}}$ and $M'=-M\sqrt{\dfrac{A(C-A)}{B(C-B)}}$ may be readily obtained.

therefore be very nearly constant. Lastly, the sine of the angle formed by the real axis of rotation and by the third principal axis will be $\frac{\sqrt{q^2+r^2}}{p}$.

If at the origin of the movement we have $q=0$ and and $r=0$, that is to say, if the real axis of rotation coincides at this instant with the third principal axis, we shall have $M=0$ and $M'=0$; q and r will be therefore always nothing, and the axis of rotation will always coincide with the third principal axis; from which it follows, that if the body begins to turn about one of the principal axes, it will continue to turn uniformly about the same axis. This remarkable property of the principal axes, has caused them to be called the principal axes of rotation; it belongs exclusively to them; for if the real axis of rotation is invariable at the surface of the body, we have $dp=0$, $dq=0$, and $dr=0$; the preceding values of these quantities therefore give

$$\frac{B-A}{C}\cdot rq=0; \quad \frac{C-B}{A}\cdot rp=0; \quad \frac{A-C}{B}\cdot pq=0.$$

In the general case where A, B, and C are unequal, two of the three quantities p, q, and r are nothing in consequence of these equations, which implies, that the real axis of rotation coincides with one of the principal axes.

If two of the three quantities A, B, and C are equal, for example, if we have $A=B$; the three preceding equations will be reduced to these $rp=0$ and $pq=0$, and they may be satisfied by supposing $p=0$. The axis of rotation is then in a plane perpendicular to the third principal axis; but we have seen, No. 27, that all the axes situated in this plane are principal axes.

Lastly, if we have at the same time $A = B = C$, the three preceding equations will be satisfied whatever may be p, q, and r, but then by No. 27, all the axes of the body are principal axes.

It follows from the above, that the principal axes alone have the property of being invariable axes of rotation; but they do not all of them possess it in the same manner. The movement of rotation about that of which the moment of inertia is between the moments of inertia of the two other axes, may be troubled in a sensible manner by the slightest cause; so that there is no stability in this movement.

That state of a system of bodies is called stable in which, when the system undergoes an indefinitely small alteration, it will vary in an indefinitely small degree by making continual oscillations about this state. This being understood, let us suppose that the real axis of rotation is at an indefinitely small distance from the third principal axis; in this case the constant quantities M and M' are indefinitely small; if n be a real quantity the values of q and r will always remain indefinitely small, and the real axis of rotation will only make oscillations of the same order about the third principal axis. But if n be imaginary, $\sin.(nt+\gamma)$ and $\cos.(nt+\gamma)$ will be changed into exponentials; consequently the expressions of q and r may augment indefinitely, and eventually cease to be indefinitely small quantities*; there is not therefore any stability

---

* By the rules of trigonometry $M.\sin.(nt+\gamma) = M . \dfrac{e^{(nt+\gamma)\sqrt{-1}} - e^{-(nt+\gamma)\sqrt{-1}}}{2\sqrt{-1}}$, in this expression e

in the movement of rotation of a body about the third principal axis*. The value of n is real if C is the

---

represents the number of which the hyperbolical logarithm is unity. If n is an impossible quantity it may be supposed equal to $m\sqrt{-1}$; let this value be substituted for it in the above equation, and we shall have $q = M.\sin.(mt\sqrt{-1}$
$+\gamma) = M . \dfrac{e^{-mt+\gamma\sqrt{-1}} - e^{mt-\gamma\sqrt{-1}}}{2\sqrt{-1}}$. As the quantity mt is real, the value of q may increase as that of t increases, until it ceases to be indefinitely small. It may be shewn in a similar manner that if t increase r will also increase, and at length cease to be indefinitely small.

* In the equations
$$p'^2 + q'^2 + r'^2 = k^2,$$
$$AB.p'^2 + BC.q'^2 + AC.r'^2 = H^2,$$
let the values of p', q', and r' be substituted, then if the second, after having divided both its members by AB, be subtracted from the first, the following will be obtained,
$$A(A-C)q^2 + B(B-C)r^2 = k^2 - \dfrac{H^2}{AB}.$$
If q and r are very small at the beginning of the movement, the constant quantity $k^2 - \dfrac{H^2}{AB}$ which may be represented by L, is very small at that time; the quantities q^2 and r^2 will therefore, if the difference $A-C$ and $B-C$ have the same sign, always remain very small and have for their respective limits $\dfrac{L}{A(A-C)}$ and $\dfrac{L}{B(B-C)}$.

If the differences $A-C$ and $B-C$ have not the same sign and the constant quantity L is supposed very small, the above equation may have place although the values of q and and r increase indefinitely.

greatest or the least of the three quantities A, B, and C, for then the product $(C-A).(C-B)$ is positive; but this product is negative if C is between A and B, and in this case n is imaginary; thus the movement of rotation is stable about the two principal axes of which the moments of inertia are the greatest and the least; but not so about the other principal axis.

In order to determine the position of the principal axes in space, let us suppose the third principal axis very nearly perpendicular to the plane of x' and y', so that θ may be a very small quantity of which the square can be neglected. We shall have by No. 26,
$$d\theta - d\psi = p\,dt;$$
which gives from integration
$$\psi = \varphi - pt - \varepsilon,$$
ε being a constant quantity. If we afterwards make
$$\sin.\theta.\sin.\varphi = s; \quad \sin.\theta.\cos.\varphi = u;$$
the values of q and r of No. 26, will give by extracting $d\psi$,
$$\frac{ds}{dt} - pu = r; \quad \frac{du}{dt} + ps = -q;$$
and by integration
$$s = \beta.\sin.(pt+\lambda) - \frac{AM}{Cp}.\sin.(nt+\gamma);$$
$$u = \beta.\cos.(pt+\lambda) - \frac{BM'}{Cp}.\cos.(nt+\gamma)*;$$

---

* By differentiation and substitution the equations
$$\frac{ds}{dt} - pu = r, \quad \frac{du}{dt} + ps = -q,$$
will become

β and λ being two new constant quantities; the problem is thus completely resolved, because the values of s and u give the angles θ and φ in a function of the time, and ψ is determined in a function of φ and t. If β is nothing the plane of x' and y' becomes the invariable plane to which we have referred in the preceding number the angles θ, φ, and ψ.

31. If the body is free, the analysis of the preceding numbers will give its movement about its centre of gravity; if the solid is forced to move about a fixed point, it will always shew its movement about this point. It remains for us to consider the movement of a solid subjected to turn about a fixed axis.

Let us suppose that x is the axis which we shall imagine to be horizontal: in this case the last of the equations (B) of No. 25 will be sufficient to determine the movement of the body. Let us suppose also, that the axis of y' is horizontal and that the axis of z' is vertical and directed towards the centre of the earth; let us conceive lastly, that the plane which passes by the axes of y' and z', passes through the centre of gravity of the body, and also that an axis passes constantly

$$\frac{d^2s}{dt^2} + p^2s - \frac{dr}{dt} + pq = 0,$$

$$\frac{d^2u}{dt^2} + p^2u + \frac{dq}{dt} + pr = 0.$$

These equations may be readily integrated. See No. 623 of the Traité du Calcul Differentiel et du Calcul Integral of Lacroix, where a general formula is given for equations of this description.

through this centre and the origin of the co-ordinates. Let θ be the angle which this new axis makes with that of x'; if we name the co-ordinates referred to this new axis, y'' and z'' we shall have*

$$y' = y''.\cos.\theta + z''.\sin.\theta; \quad z' = z''.\cos.\theta - y''.\sin.\theta;$$

from which may be obtained

$$S.\frac{y'dz' - z'dy'}{dt}.dm = -\frac{d\theta}{dt}.S.dm.(y''^2 + z''^2).$$

$S.dm.(y''^2 + z''^2)$ is the moment of inertia of the body relative to the axis of x'; let C be this moment. The last of the equations (B) of No. 25 will give

$$-C.\frac{d^2\theta}{dt^2} = \frac{dN''}{dt}.$$

Let us suppose that the body is solicited only by the force of gravity: the values of P and Q of No. 25 will be nothing and R will be constant, which gives

$$\frac{dN''}{dt} = S.Ry'.dm = R.\cos.\theta.S.y''.dm + R.\sin.\theta.S.z''.dm.$$

The axis of z'' passing through the centre of gravity of the body, we have $S.y''.dm = 0$; moreover if h represents the distance of the centre of gravity of the body

---

* By referring to figure 17, and notes page 169, A may be supposed to be the centre of the co-ordinates and point through which the horizontal axis x' passes, AY the vertical axis of z', AX the horizontal axis of y', AY, that of z'' which passes through the centre of gravity of the body and makes the angle θ with z', and AX, the axis of y''. In this case the values of y' and z' may be found in the terms of y'' and z'', in a manner similar to that in which the values of x and y were found in the terms of x, and y, in the above mentioned number.

from the axis of x', we shall have $S.x''.dm = mh$, m being the whole mass of the body; we shall therefore have

$$\frac{dN''}{dt} = mh.R.\sin.\theta;$$

and consequently

$$\frac{d^2\theta}{dt^2} = \frac{-mh.R.\sin.\theta}{C}.$$

Let us now consider a second body, all the parts of which are united in one sole point at the distance l from the axis of x'; we shall have relative to this body $C = m'l^2$, m' being its mass; moreover h will be equal to l; by equating

$$\frac{d^2\theta}{dt^2} = \frac{-R}{l}.\sin.\theta.$$

These two bodies will therefore have exactly the same movement of oscillation, if their initial angular velocities, when their centres of gravity are in the vertical, are the same, and we have $l = \frac{C}{mh}$.

The second body just noticed is the simple pendulum, of which we have considered the oscillations at No. 11; we are therefore always able to assign by this formula the length l of the simple pendulum, the oscillations of which are isochronous to those of the solid which has here been considered and which forms a compound pendulum. It is thus that the length of the simple pendulum which oscillates seconds, is determined by observations made upon compound pendulums.

CHAP. VIII.

Of the motion of fluids.

32. WE shall make the laws of the motion of fluids depend upon those of their equilibrium, in a similar manner to that by which we have in Chap. 5, deduced the laws of the motion of a system of bodies from those of its equilibrium. Let us therefore resume the general equation of the equilibrium of fluids given in No. 17;

$$\delta p = \varrho . \{ P.\delta x + Q.\delta y + R.\delta z \} ;$$

the characteristic δ being only relative to the co-ordinates x, y, and z of the molecule, and independent of the time t. When the fluid is in motion, the forces which would retain its molecules in the state of equilibrium are by No. 18, dt being supposed constant,

$$P - \left(\frac{d^2x}{dt^2}\right); \quad Q - \left(\frac{d^2y}{dt^2}\right); \quad R - \left(\frac{d^2z}{dt^2}\right);$$

it is therefore necessary to substitute these forces for P, Q, and R, in the preceding equation of equilibrium. Denoting by δV the variation $P.\delta x + Q.\delta y + R.\delta z$, which we will suppose exact*; we shall have

$$\delta V - \frac{\delta p \dagger}{\varsigma} = \delta x \cdot \left(\frac{d^2 x}{dt^2}\right) + \delta y \cdot \left(\frac{d^2 y}{dt^2}\right) + \delta z \cdot \left(\frac{d^2 z}{dt^2}\right); \quad (F)$$

this equation is equivalent to three distinct equations, for the variations δx, δy, and δz being independent, we may equal their co-efficients separately to nothing‡.

---

* As this variation is exact in the cases in which the forces of attraction are directed towards centres that are either fixed or moveable, it comprehends all the forces in nature which can act upon the molecules of a fluid mass, and may therefore be regarded as always exact.

† In places where an incompressible fluid is supported at one of its sides, the value of p shews the pressure against this side in the direction of a normal to it; at those parts of the fluid mass which are free this value is nothing. When the value of p is a known function of t, x, y, and z, it will give, by being equalled to nothing, the equation of the surface of an incompressible fluid during its motion. If t is not contained in this value of p, the surface of the fluid will preserve the same form and the same position in space, on the contrary when p contains t it will change its form or position every instant.

‡ In order that the reader may have a correct idea of the corresponding variations of t, x, y, and z and the total or partial variations of a function of them, I shall suppose F a function of t, x, y, and z, and first imagine x, y, and z to vary, t remaining constant. In this case the contemporaneous values of F may be compared, which at a determinate instant answer to the different points of a system and belong to the different molecules placed at these points at the same instant.

If on the contrary x, y, and z are supposed constant and t to vary, the values of F will appertain to the different

The co-ordinates x, y, and z are functions of the primitive co-ordinates and of the time t; let a, b, and c be these primitive co-ordinates, we shall then have

$$\delta x = \left(\frac{dx}{da}\right).\delta a + \left(\frac{dx}{db}\right).\delta b + \left(\frac{dx}{dc}\right).\delta c;$$

$$\delta y = \left(\frac{dy}{da}\right).\delta a + \left(\frac{dy}{db}\right).\delta b + \left(\frac{dy}{dc}\right).\delta c;$$

$$\delta z = \left(\frac{dz}{da}\right).\delta a + \left(\frac{dz}{db}\right).\delta b + \left(\frac{dz}{dc}\right).\delta c.$$

By substituting these values in the equation (F), the co-efficients of δa, δb, and δc may be equalled separately to nothing; which will give three equations of partial differentials between the three co-ordinates x,

---

molecules which during successive instants pass by the same point which has x, y, and z for its co-ordinates.

Lastly, if we make x, y, and z to vary either partially or together and suppose t also variable; the different values of F will belong to the same molecule, and change as it passes in successive instants from one point to another in the system. If the position of the molecule is known at the commencement of the motion, the constant quantities belonging to the three equations which give the values of x, y, and z in functions of t will be known. The values of x, y, and z may therefore be found at any instant, which will give the position of the molecule at that instant.

If the time t be eliminated from the three equations given by the values of x, y, and z, two equations of the curve described by the molecule will be known. The form and position of the curve will change by passing from one molecule to another: the constant quantities in this case changing their values as the initial position of the molecule changes.

y, and z of the molecule, its primitive co-ordinates a, b, c, and the time t.

It remains for us to fulfil the conditions of the continuity of the fluid. For this purpose let us consider at the origin of the motion, a rectangular fluid parallelepiped having for its three dimensions da, db, and dc. Denoting by (ϱ) the primitive density of this molecule, its mass will be $(\varrho).da.db.dc$. Let this parallelepiped be represented by (A)*; it is easy to

---

* In figure 22 the rectangular parallelepiped A is represented, having da, db, and dc for its three edges; this parallelepiped is changed after the time t into that given in figure 23, in which from the extremities of the edge fg which is composed of the molecules that formed the edge dc, two planes gn, fo, are supposed to be drawn parallel to the plane of x and y; by the prolongation of the edges of the parallelepiped gl or (B) to these planes a new one (C) is formed equal to (B), as the parts cut off from (B) and those added to (C) respectively compensate each other. The height fg of (C), as it is independent of the molecules in da and db, is found by making c alone to vary in differentiating the value of z, it is therefore equal to $\left(\dfrac{dz}{dc}\right).dc$. In figure 24, let δqrp denote the section (ϵ) having its side δp formed by molecules of the side db, dc, and its side δq by molecules of the side da, dc of the parallelepiped (A). From δ and p the lines δm and pn are supposed to be drawn parallel to the axis of x, meeting the line qr or its continuation in m and n, and consequently forming a new parallelogram (λ) which is equal to the former (ϵ); as it has the same base δp and is between the same parallels. The value of δp is found by taking the differential of y in making a, z, and t constant, and the value of δm by taking the differential

perceive that after the time t, it will be changed into an oblique angled parallelepiped; for all the molecules primitively situated upon any side whatever of the parallelepiped *(A)*, will again be in the same plane, at least by neglecting the indefinitely small quantities of the second order: all the molecules situated upon the parallel edges of *(A)* will be upon the small right lines equal and parallel to each other. Denoting this new parallelepiped by *(B)*, and supposing that by the extremities of the edge formed by the molecules which in the parallelepiped *(A)*, composed the edge dc, we draw two planes parallel to that of x and y. By prolonging the edges of *(B)* until they meet these two planes, we shall have a new parallelepiped *(C)* contained by them, which is equal to *(B)*; for it is evident that as much as is taken from the parallelepiped *(B)* by one of the two planes, is added to it by the other. The parallelepiped *(C)* will have its two bases parallel to the plane of x and y: its height contained between its bases will be evidently equal to the differential of z taken by making c alone to vary; which gives $\left(\frac{dz}{dc}\right).dc$ for this altitude.

---

of x in supposing y and z constant. These last values multiplied together give the value of the surface of the parallelogram (λ), or that of its equal (ι); which value, when multiplied by $\left(\frac{dz}{dc}\right).dc$ the differential of z, gives the content of the parallelepiped *(C)* or *(B)*.

We shall have its base, by observing that it is equal to the section of (B) made by a plane parallel to that of x and y; let this section be denoted by (ε). The value of z will be the same with respect to the molecules of which it is formed; and we shall have

$$0 = \left(\frac{dz}{da}\right).da + \left(\frac{dz}{db}\right).db + \left(\frac{dz}{dc}\right).dc.$$

Let δp and δq be two contiguous sides of the section (ε), of which the first is formed by molecules of the side $db.dc$ of the parallelepiped (A), and the second by molecules of its side $da.dc$. If by the extremities of the side δp we suppose two right lines parallel to the axis of x, and we prolong the side of the parallelogram (ε) parallel to δp until it meets these lines; they will intercept between themselves a new parallelogram (λ) equal to (ε), the base of which will be parallel to the axis of x. The side δp being formed by molecules of the face $db.dc$, relative to which the value of z is the same; it is easy to perceive that the height of the parallelogram (λ), is the differential of y taken by supposing u, z, and t constant, which gives

$$dy = \left(\frac{dy}{db}\right).db + \left(\frac{dy}{dc}\right).dc;$$

$$0 = \left(\frac{dz}{db}\right).db + \left(\frac{dz}{dc}\right).dc;$$

from which may be obtained

$$dy = \frac{\left\{\left(\frac{dy}{db}\right).\left(\frac{dz}{dc}\right) - \left(\frac{dy}{dc}\right).\left(\frac{dz}{db}\right)\right\}.db}{\left(\frac{dz}{dc}\right)}:$$

this is the expression of the height of the parallelogram (λ). Its base is equal to the section of this parallelogram made by a plane parallel to the axis of x; this section is formed of the molecules of the parallelepiped

(*A*) by relation to which *z* and *y* are constant, its length is therefore equal to the differential of *x* taken by supposing *z*, *y*, and *t* constant, which gives the three equations

$$dx = \left(\frac{dx}{da}\right).da + \left(\frac{dx}{db}\right).db + \left(\frac{dx}{dc}\right).dc ;$$

$$0 = \left(\frac{dy}{da}\right).da + \left(\frac{dy}{db}\right).db + \left(\frac{dy}{dc}\right).dc ;$$

$$0 = \left(\frac{dz}{da}\right).da + \left(\frac{dz}{db}\right).db + \left(\frac{dz}{dc}\right).dc.$$

Suppose for abridgment

$$\beta = \left(\frac{dx}{da}\right).\left(\frac{dy}{db}\right).\left(\frac{dz}{dc}\right) - \left(\frac{dx}{da}\right).\left(\frac{dy}{dc}\right).\left(\frac{dz}{db}\right)$$
$$+ \left(\frac{dx}{db}\right).\left(\frac{dy}{dc}\right).\left(\frac{dz}{da}\right) - \left(\frac{dx}{db}\right).\left(\frac{dy}{da}\right).\left(\frac{dz}{dc}\right)$$
$$+ \left(\frac{dx}{dc}\right).\left(\frac{dy}{da}\right).\left(\frac{dz}{db}\right) - \left(\frac{dx}{dc}\right).\left(\frac{dy}{db}\right).\left(\frac{dz}{da}\right) ;$$

we shall have

$$dx = \frac{\beta.da}{\left(\frac{dy}{db}\right).\left(\frac{dz}{dc}\right) - \left(\frac{dy}{dc}\right).\left(\frac{dz}{db}\right)} ;$$

this is the expression of the base of the parallelogram (*λ*); the surface of this parallelogram will therefore be $\frac{\beta.da.db}{\left(\frac{dz}{dc}\right)}$. This quantity also expresses the surface of the parallelogram (*ε*), if we mulitiply it by $\left(\frac{dz}{dc}\right).dc$ we shall have *β.da.db.dc* for the magnitude of the parallelepipeds (*C*) and (*B*). Let *ϱ* be the density of the parallelepiped (*A*) after the time *t*; then its mass will be represented by *ϱ.β.da.db.dc*, which being equalled with the first mass (*ϱ*)*.da.db.dc*, gives

$$\rho\beta = (\varrho) ; \qquad (G)$$

for the equation relative to the continuity of the fluid.

33. We may give to the equations *(F)* and *(G)* another form more convenient for use in certain circumstances. Let u, v, and v, be the respective velocities of a fluid molecule parallel to the axis of x, y, and z: we shall have

$$\left(\frac{dx}{dt}\right)=u;\quad \left(\frac{dy}{dt}\right)=v;\quad \left(\frac{dz}{dt}\right)=\mathrm{v}.$$

By differentiating these equations, and regarding u, v, and v as functions of the co-ordinates x, y, and z of the molecule, and of the time t; we shall have

$$\left(\frac{d^2x}{dt^2}\right)=\left(\frac{du}{dt}\right)+u.\left(\frac{du}{dx}\right)+v.\left(\frac{du}{dy}\right)+\mathrm{v}.\left(\frac{du}{dz}\right);$$

$$\left(\frac{d^2y}{dt^2}\right)=\left(\frac{dv}{dt}\right)+u.\left(\frac{dv}{dx}\right)+v.\left(\frac{dv}{dy}\right)+\mathrm{v}.\left(\frac{dv}{dz}\right);$$

$$\left(\frac{d^2z}{dt^2}\right)=\left(\frac{d\mathrm{v}}{dt}\right)+u.\left(\frac{d\mathrm{v}}{dx}\right)+v.\left(\frac{d\mathrm{v}}{dy}\right)+\mathrm{v}.\left(\frac{d\mathrm{v}}{dz}\right).$$

The equation *(F)* of the preceding number will then become

$$\delta V - \frac{\delta p}{\varrho} = \delta x . \left\{\left(\frac{du}{dt}\right)+u.\left(\frac{du}{dx}\right)+v.\left(\frac{du}{dy}\right)\right.$$
$$\left. +\mathrm{v}.\left(\frac{du}{dz}\right)\right\}$$
$$+\delta y . \left\{\left(\frac{dv}{dt}\right)+u.\left(\frac{dv}{dx}\right)+v.\left(\frac{dv}{dy}\right)\right.$$
$$\left. +\mathrm{v}.\left(\frac{dv}{dz}\right)\right\} \quad (H)$$
$$+\delta z . \left\{\left(\frac{d\mathrm{v}}{dt}\right)+u.\left(\frac{d\mathrm{v}}{dx}\right)+v.\left(\frac{d\mathrm{v}}{dy}\right)\right.$$
$$\left. +\mathrm{v}.\left(\frac{d\mathrm{v}}{dz}\right)\right\}.$$

In order to have the equation relative to the continuity of a fluid, let us suppose that in the value of β of the preceding number, a, b, and c may be equal to x, y, and z, and that x, y, and z may be equal respectively to $x+udt$, $y+vdt$, and $z+vdt$, which is equivalent to taking the first co-ordinates a, b, and c indefinitely near to x, y, and z; we shall then have

$$\beta = 1 + dt \cdot \left\{ \left(\frac{du}{dx}\right) + \left(\frac{dv}{dy}\right) + \left(\frac{dv}{dz}\right) \right\};$$

the equation (G) becomes

$$\rho dt \cdot \left\{ \left(\frac{du}{dx}\right) + \left(\frac{dv}{dy}\right) + \left(\frac{dv}{dz}\right) \right\} + \rho - (\rho) = 0.$$

If we consider ρ as a function of x, y, z, and t, we shall have

$$(\rho) = \rho - dt \cdot \left(\frac{d\rho}{dt}\right) - udt \cdot \left(\frac{d\rho}{dx}\right) - vdt \cdot \left(\frac{d\rho}{dy}\right) - vdt \cdot \left(\frac{d\rho}{dz}\right)$$

the preceding equation will therefore become

$$0 = \left(\frac{d\rho}{dt}\right) + \left(\frac{d\cdot\rho u}{dx}\right) + \left(\frac{d\cdot\rho v}{dy}\right) + \left(\frac{d\cdot\rho v}{dz}\right)*; \qquad (K)$$

this is the equation relative to the continuity of the fluid, and it is easy to perceive, that it is the differential of the equation (G) of the preceding number, taken relative to the time t†.

---

* The equation

$$0 = \left(\frac{d\rho}{dt}\right) + \left(\frac{d\cdot\rho u}{dx}\right) + \left(\frac{d\cdot\rho v}{dy}\right) + \left(\frac{d\cdot\rho v}{dz}\right).$$

is equivalent to the following

$$0 = \left(\frac{d\rho}{dt}\right) + \rho \cdot \left(\frac{du}{dx}\right) + u \cdot \left(\frac{d\rho}{dx}\right) + \rho \cdot \left(\frac{dv}{dy}\right) + v \cdot \left(\frac{d\rho}{dy}\right) + \&c.$$

which is what the equation (G) becomes when the values of β and (ρ) are substituted.

† If the fluid be incompressible, as the mass the density and the magnitude of each molecule of the fluid will remain invariable, the equation (K), by equalling separately the

The equation (H) is susceptible of integration in a very extensive case, that is when $u\delta x + v\delta y + v\delta z$ is an

---

variations of the density and the mass to nothing, will give the two following

$$\left(\frac{d\varrho}{dt}\right) + \left(\frac{d\varrho}{dx}\right)u + \left(\frac{d\varrho}{dy}\right)v + \left(\frac{d\varrho}{dz}\right)v = 0$$

$$\left(\frac{du}{dx}\right) + \left(\frac{dv}{dy}\right) + \left(\frac{dv}{dz}\right) = 0.$$

By joining these equations to the three given by that of (H) we shall have five which will enable us to determine the unknown quantities ϱ, p, u, v, and v in functions of t, x, y, and z.

If the incompressible fluid is homogeneous the density ϱ will be a constant quantity; in this case we shall have only the second of the above equations and the three given by that of (H) to determine the four unknown quantities p, u, v, and v.

If the fluid is elastic we shall have the equation (K) and the three given by that of (H): if the temperature be the same throughout the mass, the density will be as the pressure, which gives $p = k\varrho$; therefore there will be only four unknown quantities which the four equations above mentioned are sufficient to dicover. If the temperature be variable and a given function of the time, the quantity k will be a function of these variables, consequently the before mentioned equations will be sufficient to determine the values of p, u, v, and v.

It appears from the above that we shall have in every case as many equations as there are unknown quantities in the problem. But as these are equations of partial differentiations of t, x, y, and z, they have at present resisted every attempt to integrate them. In some instances they have been simplified and integrated by particular suppositions,

exact variation of x, y, and z, ρ being also any function whatever of the pressure p. If therefore $\delta\varphi$ represent this variation; the equation (H) will give*

$$\delta V - \frac{\delta p}{\rho} = \delta.\left(\frac{d\varphi}{dt}\right) + \tfrac{1}{2}.\delta.\left\{\left(\frac{d\varphi}{dx}\right)^2 + \left(\frac{d\varphi}{dy}\right)^2 + \left(\frac{d\varphi}{dz}\right)^2\right\}$$

from which we may obtain by integrating it with respect to δ,

$$V - \int\frac{\delta p}{\rho} = \left(\frac{d\varphi}{dt}\right) + \tfrac{1}{2}.\left\{\left(\frac{d\varphi}{dx}\right)^2 + \left(\frac{d\varphi}{dy}\right)^2 + \left(\frac{d\varphi}{dz}\right)^2\right\}.$$

It is necessary to add a constant quantity which is a function of t to this integral, but we may suppose that this quantity is contained in the function φ. This last function gives the velocities of the fluid molecules parallel to the axes of x, y, and z: for we have

$$u = \left(\frac{d\varphi}{dx}\right); \quad v = \left(\frac{d\varphi}{dy}\right); \quad \mathrm{v} = \left(\frac{d\varphi}{dz}\right);$$

The equation (K) relative to the continuity of the fluid becomes

---

but even then the greatest difficulty has attended the determination of the arbitrary constant quantities which depend upon the state of the fluid at the commencement of its motion.

* That the equation (H) gives

$$\delta V - \frac{\delta p}{\rho} = \delta.\left(\frac{d\varphi}{dt}\right) + \tfrac{1}{2}\delta.\left\{\left(\frac{d\varphi}{dx}\right)^2 + \left(\frac{d\varphi}{dy}\right)^2 + \left(\frac{d\varphi}{dz}\right)^2\right\}$$

may be rendered evident from considering, for instance, $\tfrac{1}{2}\delta.\left(\frac{d\varphi}{dx}\right)^2$ which is equivalent to $\left(\frac{d\varphi}{dx}\right).\delta.\left(\frac{d\varphi}{dx}\right)$ but $\frac{d\varphi}{dx} = u$ and $\delta.\left(\frac{d\varphi}{dx}\right) = \left(\frac{du}{dx}\right).\delta x + \left(\frac{dv}{dx}\right).\delta y + \left(\frac{d\mathrm{v}}{dx}\right).\delta z$, therefore $\tfrac{1}{2}.\delta.\left(\frac{d\varphi}{dx}\right)^2 = u.\left(\frac{du}{dx}\right).\delta x + u.\left(\frac{dv}{dx}\right).\delta y + u.\left(\frac{d\mathrm{v}}{dx}\right).\delta z.$

$$0 = \left(\frac{d\rho}{dt}\right) + \left(\frac{d\rho}{dx}\right) \cdot \left(\frac{d\varphi}{dx}\right) + \left(\frac{d\rho}{dy}\right) \cdot \left(\frac{d\varphi}{dy}\right)$$
$$+ \left(\frac{d\rho}{dz}\right) \cdot \left(\frac{d\varphi}{dz}\right) + \rho \cdot \left\{\left(\frac{d^2\varphi}{dx^2}\right) + \left(\frac{d^2\varphi}{dy^2}\right) + \left(\frac{d^2\varphi}{dz^2}\right)\right\};$$

thus we have relative to homogeneous fluids

$$0 = \left(\frac{d^2\varphi}{dx^2}\right) + \left(\frac{d^2\varphi}{dy^2}\right) + \left(\frac{d^2\varphi}{dz^2}\right)*.$$

It may be observed that the function $u.\delta x + v.\delta y + \mathrm{v}.\delta z$ is an exact variation of x, y, and z at all times, if it be during one instant. Let us suppose that at any instant whatever, it is equal to $\delta\varphi$; in the following instant we shall have

$$\delta\dot\varphi + dt . \left\{\left(\frac{du}{dt}\right).\delta x + \left(\frac{dv}{dt}\right).\delta y + \left(\frac{d\mathrm{v}}{dt}\right).\delta z\right\};$$

it will therefore be an exact variation at this instant, if $\left(\frac{du}{dt}\right).\delta x + \left(\frac{dv}{dt}\right).\delta y + \left(\frac{d\mathrm{v}}{dt}\right).\delta z$ is an exact variation at the first instant; but the equation *(H)* gives at this instant $\left(\frac{du}{dt}\right).\delta x + \left(\frac{dv}{dt}\right).\delta y + \left(\frac{d\mathrm{v}}{dt}\right).\delta z = \delta V - \frac{1}{2}.\delta.$
$\left\{\left(\frac{d\varphi}{dx}\right)^2 + \left(\frac{d\varphi}{dy}\right)^2 + \left(\frac{d\varphi}{dz}\right)^2\right\} - \frac{\delta p}{\rho}$; the first member of this equation is consequently an exact variation in x,

---

* The integration of the equation

$$0 = \left(\frac{d^2\varphi}{dx^2}\right) + \left(\frac{d^2\varphi}{dy^2}\right) + \left(\frac{d^2\varphi}{dz^2}\right),$$

which presented the greatest difficulties has been fortunately accomplished by Marc Antoine Parseval, a French mathematician. Vide the eighth Cahier of the Journal de l'Ecole Polytechnique.

y, and z; thus if the function $u.\delta x + v.\delta y + \mathrm{v}.\delta z$ be an exact variation one instant, it will also be one in the next, it is therefore an exact variation at all times.

When the motions are very small, we may neglect the squares and the products of u, v, and v; the equation (H) then becomes

$$\delta V - \frac{\delta p}{\rho} = \left(\frac{du}{dt}\right).\delta x + \left(\frac{dv}{dt}\right).\delta y + \left(\frac{d\mathrm{v}}{dt}\right).\delta z;$$

therefore in this case $u.\delta x + v.\delta y + \mathrm{v}.\delta z$ is an exact variation, if, as we have supposed, p be a function of ρ; by naming this differential $\delta\varphi$, we shall have

$$V - \int\frac{\delta p}{\rho} = \left(\frac{d\varphi}{dt}\right)*;$$

and if the fluid be homogeneous, the equation of continuity will become

$$0 = \left(\frac{d^2\varphi}{dx^2}\right) + \left(\frac{d^2\varphi}{dy^2}\right) + \left(\frac{d^2\varphi}{dz^2}\right).$$

These two equations contain the whole of the theory of very small undulations of homogeneous fluids.

34. Let us consider an homogeneous fluid mass which has an uniform movement of rotation about the

---

* In the case of the very small undulations of an homogeneous incompressible heavy fluid, such as water, $\int\frac{dp}{\rho} = \frac{p}{\rho}$; if the axis of z be supposed in the direction of gravity, at its surface the equation $V - \int\frac{dp}{\rho} = \left(\frac{d\varphi}{dt}\right)$ is changed into the following $gz = \left(\frac{d\varphi}{dt}\right)$: g representing the constant force of gravity.

axis of x. Let n represent the angular velocity of rotation at a distance from the axis which we will take for the unity of distance; we shall then have $v = -nz$; $v = ny^*$; the equation (H) of the preceding number will also become

$$\frac{\delta p}{\rho} = \delta V + n^2 . \{y \delta y + z \delta z\};$$

which equation is possible because its two members are exact differentials. The equation (K) of the same number in like manner will become

$$0 = dt . \left(\frac{d\rho}{dt}\right) + u . dt . \left(\frac{d\rho}{dx}\right) + v . dt . \left(\frac{d\rho}{dy}\right) + \mathrm{v} . dt . \left(\frac{d\rho}{dz}\right);$$

---

* In figure 25 let A represent the origin of the co-ordinates, AY the axis of y, AZ that of z, and AD the projection upon the plane yz of a line drawn from a molecule in the fluid mass perpendicular to the axis of x; from D draw the line DZ perpendicular to the axis AZ, then $DZ = y$, $AZ = z$, and $AD = \sqrt{y^2 + z^2}$. Let a line DE be drawn from D perpendicular to AD, and from any point E in it draw a line EF perpendicular to DZ; then if DE represent the velocity of the molecule in the direction perpendicular to AD, it may be supposed to be resolved into two others DF and FE in the respective directions of y and z. As the velocity at the distance represented by unity from the axis of x is n, that at the distance $\sqrt{y^2 + z^2}$ is $n\sqrt{y^2+z^2} = DE$. From the similarity of the right angled triangles AZD and DFE

$$\sqrt{y^2 + z^2} : z :: n\sqrt{y^2 + z^2} : DF = nz;$$

but $DF = nz$ is the velocity in the direction of the axis y, and ought to be taken negatively as it tends to diminish that

and it is evident that this equation is satisfied if the fluid mass be homogeneous*. The equations of the motion of fluids are then therefore satisfied, and consequently this movement is possible.

The centrifugal force at the distance $\sqrt{y^2+z^2}$ from the axis of rotation, is equal to the square of the velocity $n^2.(y^2+z^2)$ divided by this distance; the function $n^2.(y\delta y+z\delta z)$ is consequently the product of the centrifugal force by the element of its direction†; therefore by comparing the preceding equation of the movement

---

axis. Again $\sqrt{y^2+z^2} : y :: n\sqrt{y^2+z^2} : FE = ny$, or the velocity of the molecule in the direction of the axis z.

* That the equation (H) is reduced to the value given in this number appears evident from considering, that all the terms in the second member vanish except $v.\left(\dfrac{dv}{dz}\right).\delta y = -n^2 y\delta y$ and $v.\left(\dfrac{dv}{dy}\right).\delta z = -n^2 z\delta z$

In the equation (K) the respective values of $\rho\left(\dfrac{dv}{dx}\right)$, $\rho.\left(\dfrac{dv}{dy}\right)$ and $\rho.\left(\dfrac{dv}{dz}\right)$, in this case, are evidently equal to nothing: also the partial differentiations of ρ respectively vanish if the fluid is homogeneous.

† The centrifugal force at the distance $\sqrt{y^2+z^2}$ from the axis of rotation is equal to $\dfrac{n^2.(y^2+z^2)}{\sqrt{y^2+z^2}}$ or $n^2.\sqrt{y^2+z^2}$; this value multiplied into the element of its direction or $\dfrac{y\delta y+z\delta z}{\sqrt{y^2+z^2}}$ gives $n^2.(y\delta y+z\delta z)$.

of a fluid, with the general equation of the equilibrium of fluids given in No. 17; we may perceive that the conditions of movement of which it treats, reduce themselves to those of the equilibrium of a fluid mass solicited by the same forces, and by the centrifugal force due to the movement of rotation: which is otherways evident*.

If the exterior surface of a fluid mass is free, we shall have $\delta p = 0$ at this surface, and consequently
$$0 = \delta V + n^2 . (y \delta y + z \delta z);$$
from which it follows, that the resultant of all the forces which act upon each molecule of the exterior surface, should be perpendicular to this surface; it ought also to be directed towards the interior of the fluid mass. If these conditions be fulfilled an homogeneous fluid mass will be in equilibrio, supposing at the same time, that it covers a solid body of any figure whatever.

The case which we have examined is one of those in which the variation $u.\delta x + v.\delta y + \text{v}.\delta z$ is not exact: for

---

* The general equation given in this number, when the value of δV is substituted, becomes
$$\frac{\delta p}{\rho} = P.\delta x + Q.\delta y + R.\delta z + n^2.(y\delta y + z\delta z);$$
which is independent of the time, and has $n^2.(y\delta y + z\delta z)$ for the value of the centrifugal force multiplied into the element of its direction; it is therefore evident that the conditions of movement are, in this case, the same as those of the equilibrium of a fluid mass solicited by the same forces and by the centrifugal force arising from the rotatory motion of the mass. See notes page 134.

this variation becomes $-n.(z\delta y-y\delta z)$; therefore in the theory of the flux and reflux of the sea, we cannot suppose that the variation concerned is exact, because it is not in the very simple case in which the sea has no other movement than that of rotation, which is common to it and the earth.

35. Let us now determine the oscillations of a fluid mass covering a spheroid possessed of a movement of rotation nt about the axis of x; supposing it to be very little altered from the state of equilibrium by the action of very small forces. At the beginning of the movement, let r be the distance of a fluid molecule from the centre of gravity of the spheroid that it covers, which we will suppose immoveable, let θ be the angle that the radius r forms with the axis of x, and ϖ the angle which the plane that passes by the axis of x and this radius forms with the plane of x and y. Let us suppose that after the time t the radius r is changed into $r+\alpha s$, that the angle θ is changed into $\theta+\alpha u$, and that the angle ϖ is changed into $nt+\varpi+\alpha v$; αs, αu, and αv being very small quantities of which we may neglect the squares and the products; we shall then have*

---

* In figure 18, let C represent the origin of the co-ordinates at the centre of gravity of the spheroid, CB, CF, and CE the respective axes of x, y, and z, CA the distance of a fluid molecule from C, AD a perpendicular let fall from the molecule to the plane of FCB or xy, DB and consequently AB, perpendiculars drawn from D and A to the axis of x, then $CA=r+\alpha s$, $CB=x$, $BD=y$, and $AD=z$, also the angle ACB made by the radius and axis of $x=\theta+\alpha u$, and

$$x = (r+\alpha s).\cos.(\theta+\alpha u);$$
$$y = (r+\alpha s).\sin.(\theta+\alpha u).\cos.(nt+\varpi+\alpha v);$$
$$z = (r+\alpha s).\sin.(\theta+\alpha u).\sin.(nt+\varpi+\alpha v).$$

If we substitute these values in the equation *(F)* of No. 32, we shall have by neglecting the square of α,

$$\alpha r^2.\delta\theta.\left\{\left(\frac{d^2u}{dt^2}\right) - 2n.\sin.\theta.\cos.\theta.\left(\frac{dv}{dt}\right)\right\}$$
$$+\alpha r^2.\delta\varpi\left\{\sin.^2\theta.\left(\frac{d^2v}{dt^2}\right) + 2n.\sin.\theta.\cos.\theta.\left(\frac{du}{dt}\right)\right.$$
$$\left.+\frac{2n.\sin.^2\theta}{r}\left(\frac{ds}{dt}\right)\right\} + \alpha\,\delta r\cdot\left\{\left(\frac{d^2s}{dt^2}\right) - 2nr.\right. \quad (L)$$
$$\left.\sin.^2\theta.\left(\frac{dv}{dt}\right)\right\} = \frac{n^2}{2}.\delta.\{(r+\alpha s).\sin.(\theta+\alpha u)\}^2$$
$$+\delta V - \frac{\delta p}{\rho}.$$

At the exterior surface of the fluid $\delta p = 0$; also in the state of equilibrium

$$0 = \frac{n^2}{2}.\delta.\{(r+\alpha s).\sin.(\theta+\alpha u)\}^2 + (\delta V);$$

(δV) being the value of δV which belongs to this state.

---

the angle *ABD* or the inclination of the planes *ACB* and $xy = nt+\varpi+\alpha v$. By trigonometry, in the right angled triangle *ACB* we have
rad.(1) : cos.$(\theta+\alpha u)$: : $r+\alpha s$: $x = (r+\alpha s)\cos.(\theta+\alpha u)$,
and
rad.(1) : sin.$(\theta+\alpha u)$: : $r+\alpha s$: $AB = (r+\alpha s).\sin.(\theta+\alpha u)$;
also in the right angled triangle *ADB* we have
rad.(1) : cos.$(nt+\varpi+\alpha v)$: : $(r+\alpha s)\sin.(\theta+\alpha u)$: $y = (r+\alpha s)\sin.(\theta+\alpha u)\cos.(nt+\varpi+\alpha v)$ and
rad.(1) : sin.$(nt+\varpi+\alpha v)$: : $(r+\alpha s)\sin.(\theta+\alpha u)$: $z = (r+\alpha s).\sin.(\theta+\alpha u).\sin.(nt+\varpi+\alpha v).$

Let us suppose the sea to be the fluid treated upon; the variation (δV) will be the product of the gravity multiplied by the element of its direction. Let g represent the force of gravity, and αy the elevation of a molecule of water at the surface above the surface of equilibrium, which surface we shall regard as the true level (niveau) of the sea. The variation (δV) in the state of movement will by this elevation be increased by the quantity $-\alpha g.\delta y$, because the force of gravity acts very nearly in the direction of the αy's and towards their origin. Lastly denoting by $\alpha \delta V'$ the part of δV relative to the new forces which in the state of movement solicit the molecule, and depend either upon the changes which the attractions of the spheroid and the fluid experience from this state, or from foreign attractions; we shall have at the surface

$$\delta V = (\delta V) - \alpha g.\delta y + \alpha.\delta V'.$$

The variation $\frac{n^2}{2}.\delta.\{(r+\alpha s).\sin.(\theta+\alpha u)\}^2$ is increased by the quantity $\alpha n^2.\delta y.r.\sin.^2\theta$*, in consequence of the height of the molecule of water above the surface of the sea; but this quantity may be neglected relative to the term $-\alpha g.\delta y$, because the ratio $\frac{n^2 r}{g}$ of the centrifugal

---

* The quantity $\alpha n^2.\delta y.r.\sin.^2\theta$ may be obtained from the variation $\frac{n^2}{2}.\delta.\{(r+\alpha s).\sin.(\theta+\alpha u)\}^2$ by differentiating that variation with respect to r, neglecting the quantities αs and αu, which are relative to time, and supposing that δr is equal to $\alpha \delta y$.

force at the equator to gravity, is a very small fraction equal to $\frac{1}{289}$*. Lastly, the radius r is very nearly constant at the surface of the sea, because it differs very little from a spherical surface; we may therefore suppose δr equal to nothing. The equation L thus becomes at the surface of the sea

$$r^2.\delta\theta.\left\{\left(\frac{d^2u}{dt^2}\right)-2n.\sin.\theta.\cos.\theta.\left(\frac{dv}{dt}\right)\right\}$$
$$+r^2.\delta\varpi.\left\{\sin.^2\theta.\left(\frac{d^2v}{dt^2}\right)+2n.\sin.\theta.\cos.\theta.\left(\frac{du}{dt}\right)+2n.\sin.^2\theta.\left(\frac{ds}{dt}\right)\right\}=-g.\delta y+\delta V';$$

the variations δy and $\delta V'$ being relative to the two variables θ and ϖ.

Let us now consider the equation relative to the continuity of the fluid. For which purpose, we may suppose at the origin of the movement a rectangular parallelepiped, of which the altitude is dr, the breadth $rd\varpi.\sin.\theta$ and the length $rd\theta$. Let r', θ', and ϖ' represent the values of r, θ, and ϖ after the time t. By following the reasoning of No. 32, we shall find that after this time, the volume of the fluid molecule is

---

* $\alpha n^2.\delta y.r.\sin.^2\theta$ has the same ratio to $-\alpha g.\delta y$ as $\frac{n^2.r.\sin.^2\theta}{g}$ has to -1, but the centrifugal force at the equator is $\frac{n^2r^2}{r}$ or n^2r, and $\frac{n^2r}{g}$ is nearly equal to $\frac{1}{289}$ therefore $\frac{n^2.r.\sin.^2\theta}{g}$ may be neglected when compared with -1.

equal to a rectangular parallelepiped of which the ...itude is $\left(\dfrac{dr'}{dr}\right).dr$ and the breadth

$$r'.\sin.\theta'.\left\{\left(\dfrac{d\varpi'}{d\varpi}\right).d\varpi+\left(\dfrac{d\varpi'}{dr}\right).dr\right\},$$

by extracting dr by means of the equation

$$0=\left(\dfrac{dr'}{d\varpi}\right).d\varpi+\left(\dfrac{dr'}{dr}\right).dr\,;$$

lastly its length is

$$r'.\left\{\left(\dfrac{d\theta'}{dr}\right).dr+\left(\dfrac{d\theta'}{d\theta}\right).d\theta+\left(\dfrac{d\theta'}{d\varpi}\right).d\varpi\right\}$$

by extracting dr and $d\varpi$, by means of the equations

$$0=\left(\dfrac{dr'}{dr}\right).dr+\left(\dfrac{dr'}{d\theta}\right).d\theta+\left(\dfrac{dr'}{d\varpi}\right).d\varpi\,;$$

$$0=\left(\dfrac{d\varpi'}{dr}\right).dr+\left(\dfrac{d\varpi'}{d\theta}\right).d\theta+\left(\dfrac{d\varpi'}{d\varpi}\right).d\varpi.$$

Supposing therefore

$$\beta'=\left(\dfrac{dr'}{dr}\right).\left(\dfrac{d\theta'}{d\theta}\right).\left(\dfrac{d\varpi'}{d\varpi}\right)-\left(\dfrac{dr'}{dr}\right).\left(\dfrac{d\theta'}{d\varpi}\right).\left(\dfrac{d\varpi'}{d\theta}\right)$$

$$+\left(\dfrac{dr'}{d\theta}\right).\left(\dfrac{d\theta'}{d\varpi}\right).\left(\dfrac{d\varpi'}{dr}\right)-\left(\dfrac{dr'}{d\theta}\right).\left(\dfrac{d\theta'}{dr}\right).\left(\dfrac{d\varpi'}{d\varpi}\right)$$

$$+\left(\dfrac{dr'}{d\varpi}\right).\left(\dfrac{d\theta'}{dr}\right).\left(\dfrac{d\varpi'}{d\theta}\right)-\left(\dfrac{dr'}{d\varpi}\right).\left(\dfrac{d\theta'}{d\theta}\right).\left(\dfrac{d\varpi'}{dr}\right);$$

the volume of the molecule after the time t will be $\beta'.r'^2.\sin.\theta.dr.d\theta.d\varpi$; therefore naming (ρ) the primitive density of this molecule, and ρ its density corresponding to t; we shall have by equalling the primitive expression of its mass, to its expression after the time t,

$$\rho.\beta'r'^2.\sin.\theta'=(\rho).r^2.\sin.\theta\,;$$

this is the equation of the continuity of the fluid. ✱ In the present case

$$r'=r+\alpha s\,;\quad \theta'=\theta+\alpha u\,;\quad \varpi=nt+\varpi+\alpha v\,;$$

we shall therefore have by neglecting the quantities of the order α^2,

$$\beta' = 1 + \alpha \cdot \left(\frac{ds}{dr}\right) + \alpha \cdot \left(\frac{du}{d\theta}\right) + \alpha \cdot \left(\frac{dv}{d\varpi}\right).$$

Let us suppose that after the time θ, the primitive density (ρ) of the fluid is changed into $(\rho) + \alpha\rho'$; the preceding equation relative to the continuity of the fluid will give

$$0 = r^2 \cdot \left\{ \rho' + (\rho) \cdot \left\{ \left(\frac{du}{d\theta}\right) + \left(\frac{dv}{d\varpi}\right) + \left(\frac{u \cdot \cos\theta}{\sin\theta}\right) \right\} + (\rho) \left(\frac{d \cdot r^2 s}{dr}\right).$$

36. *Let us apply these results to the oscillations of the sea. Its mass being homogeneous we have $\rho' = 0$ and consequently

$$0 = \left(\frac{d \cdot r^2 s}{dr}\right) + r^2 \cdot \left\{ \left(\frac{du}{d\theta}\right) + \left(\frac{dv}{d\varpi}\right) + \frac{u \cdot \cos\theta}{\sin\theta} \right\}.$$

Let us suppose conformably to what appears to have place in nature, that the depth of the sea is very small relative to the radius r of the terrestrial spheroid; let this depth be represented by γ, γ being a very small function of θ and ϖ which depends upon the law of this depth. If we integrate the preceding equation with respect to r, from the surface of the solid which the sea covers unto the surface of the sea; we shall

---

* As the notes necessary to elucidate this and the following number satisfactorily would from their very great length too much increase the size of the work, I shall refer the reader who is desirous of full information respecting them to the fourth book of the Mechanique Celeste, where all the equations are integrated and every particular explained in the fullest manner.

find that the value of s is equal to a function of θ, ϖ, and t independent of r, plus a very small function which will be with respect to u and v of the same small order as the function $\frac{\gamma}{r}$; but at the surface of the solid which the sea covers, when the angles θ and ϖ are changed into $\theta+\alpha u$ and $nt+\varpi+\alpha v$, it is easy to perceive, that the distance of the molecule of water contiguous to this surface from the centre of gravity of the earth, only varies by a very small quantity with respect to αu and αv, and of the same order as the products of these quantities by the eccentricity of the spheroid covered by the sea: the function independent of r which enters into the expression of s is therefore a very small quantity of the same order; so that we may generally neglect s in the expressions where n and v are concerned. The equation of the motion of the sea at its surface given in No. 35 therefore becomes

$$r^2.\delta\theta.\left\{\left(\frac{d^2u}{dt^2}\right)-2n.\sin.\theta.\cos.\theta.\left(\frac{dv}{dt}\right)\right\}$$

$$+r^2.\delta\varpi.\left\{\sin.^2\theta.\left(\frac{d^2v}{dt^2}\right)+2n.\sin.\theta.\cos.\theta.\left(\frac{du}{dt}\right)\right\}=-g.\delta y+\delta V'; \quad (M)$$

the equation (L) of the same number relative to any point whatever of the interior of the mass of fluid, gives in the state of equilibrium

$$0=\frac{n^2}{2}.\delta.\{(r+\alpha s).\sin.(\theta+\alpha u)\}^2+(\delta V)-\frac{\delta p}{\rho};$$

(δV) and (δp) being the values of δV and δp which in the state of equilibrium answer to the quantities $r+\alpha s$, $\theta+\alpha u$, and $\varpi+\alpha v$. Suppose that in the state of motion, we have

$$\delta V=(\delta V)+\alpha.\delta V'; \quad \delta p=(\delta p)+\alpha.\delta p';$$

the equation (L) will give

$$\left\{ \frac{d.\left(V' - \frac{p'}{\rho}\right)}{dr} \right\} = \left(\frac{d^2 s}{dt^2}\right) - 2nr.\sin.^2\theta.\left(\frac{dv}{dt}\right).$$

The equation (M) shews that $n.\left(\frac{dv}{dt}\right)$ is of the same order as y or s, and consequently of the order $\frac{\gamma u}{r}$; the value of the first member of this equation is therefore of the same order; thus multiplying this value by dr, and integrating it from the surface of the spheroid which the sea covers unto the surface of the sea, we shall have $V' - \frac{p'}{\rho}$ equal to a very small function of the order $\frac{\gamma s}{r}$, plus a function of θ, ϖ, and t independent of r, which we will denote by λ; having therefore regard in the equation (L) of No. 35 only to two variables θ and ϖ, it will be changed into the equation (M), with the sole difference, that the second member will be changed into $\delta\lambda$. But λ being independent of the depth at which the molecule of water which we are considering is found; if we suppose this molecule very near the surface, the equation (L) ought evidently to coincide with the equation (M); we have therefore $\delta\lambda = \delta V' - g.\delta y$, and consequently

$$\delta.\left\{ V' - \frac{p'}{\rho} \right\} = \delta V' - g.\delta y\ ;$$

the value of $\delta V'$ in the second member of this equation being relative to the surface of the sea. We shall find in the theory of the flux and reflux of the sea, that this value is nearly the same for all the molecules situated upon the same terrestrial radius, from the surface of the solid which the sea covers to the surface of the sea; we have therefore relative to all these molecules $\frac{\delta p'}{\rho} = g \cdot \delta y$; which gives p' equal to $\rho . gy$ plus a function independent of θ, ϖ, and r: but at the surface of level of the sea, the value of $\alpha p'$ is equal to the pressure of the small column αy of water which is elevated above this surface, and this pressure is equal to $\alpha \rho . gy$; we have therefore in all the interior of the fluid mass, from the surface of the spheroid which the sea covers, to the surface of level of the sea, $p' = \rho gy$; therefore any point whatever of the surface of the spheroid covered by the sea, is more pressed than in the state of equilibrium, by all the weight of the small column of water comprised between the surface of the sea and the surface of level. This excess of pressure becomes negative at the points where the surface of the sea is sunk below the surface of level.

It follows from what has been said, that if we only have regard to the variations of θ and of ϖ; the equation (L) will be changed into the equation (M), for all the interior molecules of the fluid mass. The values of u and of v relative to all the molecules of the sea situated upon the same terrestrial radius, are therefore determined by the same differential equations: therefore by supposing as we shall in the theory of the flux and reflux of the sea, that at the beginning of the motion the values of u, $\left(\frac{du}{dt}\right)$, v, $\left(\frac{dv}{dt}\right)$, were the same for all the molecules situated upon the same radius these molecules would remain upon the same radius during the oscillations of the fluid. The values r, u, and v may therefore be supposed very nearly the same upon the small part of the terrestrial radius comprised between the solid that the sea covers and the surface of the sea: therefore from integrating with relation to r the equation

$$0 = \left(\frac{d.r^2s}{dr}\right) + r^2 \cdot \left\{\left(\frac{du}{d\theta}\right) + \left(\frac{dv}{d\varpi}\right) + \frac{u.\cos.\theta}{\sin.\theta}\right\};$$

we shall have

$$0 = r^2s - (r^2s) + r^2\gamma \cdot \left\{\left(\frac{du}{d\theta}\right) + \left(\frac{dv}{d\varpi}\right) + \frac{u.\cos.\theta}{\sin.\theta}\right\};$$

(r^2s) being the value of r^2s at the surface of the spheroid covered by the sea. The function $r^2s - (r^2s)$ is very nearly equal to $r^2 \cdot \{s - (s)\} + 2r\gamma(s)$, (s) being the value of s at the surface of the spheroid; the term $2r\gamma.(s)$ may be neglected on account of the smallness of γ and (s); thus we shall have

$$r^2s - (r^2s) = r^2 \cdot \{s - (s)\}.$$

Moreover the depth of the sea corresponding to the angles $\theta + \alpha u$ and $nt + \varpi + \alpha v$ is $\gamma + \alpha \cdot \{s - (s)\}$; if we place the origin of the angles θ and $nt + \varpi$ at a point and a meridian fixed upon the surface of the earth, which may be done as we shall forthwith see; this depth will be $\gamma + \alpha u \cdot \left(\frac{d\gamma}{d\theta}\right) + \alpha v \cdot \left(\frac{d\gamma}{d\varpi}\right)$, plus the elevation αy of the fluid molecule of the surface of the sea above the surface of level; we shall therefore have

$$s - (s) = y + \alpha \cdot \left(\frac{d\gamma}{d\theta}\right) + v \cdot \left(\frac{d\gamma}{d\varpi}\right).$$

The equation relative to the continuity of the fluid consequently will become

$$y = -\left(\frac{d.\gamma u}{d\theta}\right) - \left(\frac{d.\gamma v}{d\varpi}\right) - \frac{\gamma u.\cos.\theta}{\sin.\theta}. \qquad (N)$$

It may be observed that in this equation, the angles θ and $nt + \varpi$ are reckoned relative to a point and to a meridian fixed upon the earth, and that in the equation (M) these same angles are reckoned relative to the axis of x, and to a plane which passing through this axis will have a movement of rotation about it equal to n; but this axis and this plane are not fixed at the surface of the earth, because the attraction and the pressure of the fluid which covers it ought to alter their position a little upon this surface, as well as the movement of rotation of the spheroid. But it is easy to perceive, that these alterations are to the values of αu and αv, in the ratio of the mass of the sea to that of the terrestrial spheroid; thus, in order to refer the angles θ and $nt+\varpi$ to a point and to a meridian which are invariable at the surface of this spheriod in the two equations (M) and (N); it is sufficient to alter u and v by quantities of the order $\frac{\gamma u}{r}$ and $\frac{\gamma v}{r}$, which quantities may be neglected; in these equations therefore, it may be supposed that αu and αv are the movements of the fluid in latitude and longitude.

Again, it may be observed, that the centre of gravity of the spheroid being supposed immoveable, it is necessary to transfer in a different direction to the fluid molecules the forces by which it is actuated in consequence of the re-action of the sea; but as the common centre of gravity of the spheroid and the sea does not change its situation in consequence of this re-action, it is evident that the ratio of these forces to those by which the molecules are impelled from the action of the spheroid, is of the same order as the ratio of the fluid mass to that of the spheroid, and consequently of the order $\frac{\gamma}{r}$; they may therefore be neglected in the calculation of $\delta V'$.

37. Let us consider in the same manner the motions of the atmosphere. We shall in this research neglect the variations of the heat at different latitudes and different heights, as well as all irregular causes which

agitate it, and only have regard to the regular causes which act upon it as upon the ocean. We shall consequently suppose the sea covered by an elastic fluid of an uniform temperature; we will also suppose conformably to experience, that the density of this fluid is proportional to its pressure. This supposition gives an indefinite height to the atmosphere, but it is easy to be assured, that at a very small height its density is so trifling that it may be regarded as nothing.

This being agreed upon, let s', u', and v' represent for the molecules of the atmosphere, what s, u, and v signified for the molecules of the sea; the equation (L) of No. 35 will then give

$$\alpha r^2 . \delta\theta . \left\{ \left(\frac{d^2 u'}{dt^2}\right) - 2n.\sin.\theta.\cos.\theta.\left(\frac{dv'}{dt}\right) \right\}$$

$$+ \alpha r^2 . \delta\varpi \left\{ \sin.^2\theta.\left(\frac{d^2 v'}{dt^2}\right) + 2n.\sin.\theta.\cos.\theta. \right.$$

$$\left. \left(\frac{du'}{dt}\right) + \frac{2n.\sin.^2\theta}{r}.\left(\frac{ds'}{dt}\right) \right\} + \alpha \delta r . \left\{ \left(\frac{d^2 s'}{dt^2}\right) \right.$$

$$\left. - 2nr.\sin.^2\theta.\left(\frac{dv'}{dt}\right) \right\} = \frac{n^2}{2}.\delta.\{(r+\alpha s').$$

$$\sin.(\theta+\alpha u')\}^2 + \delta V - \frac{\delta p}{\rho}.$$

Let us at present consider the atmosphere in the state of equilibrium in which s', u', and v' are nothing. The preceding equation will give by integration,

$$\frac{n^2}{2}.r^2.\sin.^2\theta + V - \int\frac{\delta p}{\rho} = \text{constant}.$$

The pressure p being conceived to be proportional to the density, we shall make $p = l.g.\rho$, g being the gravity at a determinate place, which may be supposed to be the equator, and l being a constant quantity that gives the height of the atmosphere, conceived to be of the same density throughout, as at the surface of the sea:

this height is very small when compared with the radius of the terrestrial spheroid, of which it is not the 720th part.

The integral $\int \frac{\delta p}{\varphi}$ is equal to $l.g.\log.\rho$; the preceding equation of the equilibrium of the atmosphere consequently becomes

$$lg.\log.\rho = \text{const.} + V + \frac{n^2}{2}.r^2\sin.^2\theta.$$

At the surface of the sea, the value of V is the same for a molecule of air as for the molecule of water which is contiguous to it, because the forces which solicit each molecule are the same; but the conditions of the equilibrium of the sea requires that we should have

$$V + \frac{n^2}{2}.r^2.\sin.^2\theta = \text{const.};$$

therefore at this surface ρ is constant, that is to say, the density of the lamina of air next to the sea, is throughout the same in the state of equilibrium.

If R represent the part of the radius r comprised between the centre of the spheroid and the surface of the sea, and r' the part comprised between this surface and a molecule of air elevated above it, r' will only differ by quantities nearly of the order $\frac{\left(\frac{n^2}{g}.r'\right)^2}{R}$

from the height of this molecule above the surface of the sea: we shall therefore neglect the quantities of this order. The equation between ρ and r will give

$$lg.\log.\rho = \text{const.} + V + r'.\left(\frac{dV}{dr}\right) + \frac{r'^2}{2}.\left(\frac{d^2V}{dr^2}\right) + \frac{n^2}{2}.R^2.$$

$\sin.^2\theta + n^2.Rr'.\sin.^2\theta$; the values of V, $\left(\frac{dV}{dr}\right)$ and

$\left(\frac{d^2V}{dr^2}\right)$ being relative to the surface of the sea where we have

$$\text{const.} = V + \frac{n^2}{2}.R_2.\sin.^2\theta\ ;$$

the quantity $-\left(\frac{dV}{dr}\right) - n^2 R.\sin.^2\theta$ is the gravity at this same surface; we shall denote it by g'. The function $\left(\frac{d^2V}{dr^2}\right)$ being multiplied by the very small quantity r'^2, we may determine it on the supposition that the earth is spherical, and neglect the density of the atmosphere relative to that of the earth; we shall thus have very nearly

$$-\left(\frac{dV}{dr}\right) = g = \frac{m}{R^2}\ ;$$

m denoting the mass of the earth; by equating $\left(\frac{d^2V}{dr}\right) = \frac{2m}{R^3} = \frac{2g'}{R}$; we shall therefore have $lg.\log.\rho = \text{const.} - r'g' - \frac{r'^2}{R}.g'$; from which may be obtained

$$\rho = \Pi.c^{-\frac{r'g'}{lg}.\left(1 - \frac{r'}{R}\right)}$$

c being the number of which the hyperbolical logarithm is unity, and Π being a constant quantity evidently equal to the density of the air at the surface of the sea. Let h and h' represent the respective lengths of pendulums oscillating seconds at the surface of the sea under the equator, and at the latitude of the molecule of air which has been considered: we shall have $\frac{g'}{g} = \frac{h'}{h}$, and consequently

$$-\frac{r'h'}{lh}\cdot\left(1-\frac{r'}{R}\right).$$

$$\rho = \Pi.c$$

This expression of the density of the air shews, that the laminæ of the same density are throughout equally elevated above the surface of the sea, except by the quantity $\frac{r'(h'-h)}{h}$ nearly; but in the exact calculation of the heights of mountains by the observations of the barometer, this quantity ought not to be neglected.

Let us now consider the atmosphere in the state of motion, and let us determine the oscillations of a lamina of level, or of the same density in the state of equilibrium. Let $\alpha\varphi$ be the elevation of a molecule of air above the surface of level to which it appertains in the state of equilibrium; it is evident that in consequence of this elevation, the value of δV will be augmented by the differential variation $-\alpha g.\delta\varphi$; we shall therefore have $\delta V = (\delta V) - \alpha g.\delta\varphi + \alpha\delta V'$; (δV) being the value of δV which in the state of equilibrium corresponds to the lamina of level and to the angles $\theta + \alpha u$ and $nt + \varpi + \alpha v$; $\delta V'$ being the part of δV arising from the new forces which in the state of movement agitate the atmosphere.

Let $\rho = (\rho) + \alpha\rho'$, (ρ) being the density of the lamina of surface in the state of equilibrium. If we make $\frac{l\rho'}{(\rho)} = y'$, we shall have

$$\frac{\delta p}{\rho} = \frac{lg.\delta(\rho)}{(\rho)} + \alpha g.\delta y';$$

but in the state of equilibrium

$$0 = \frac{n^2}{2}.\delta.\{(r+\alpha s).\sin.(\theta+\alpha u)\}^2 + (\delta V) - \frac{lg.\delta(\rho)}{(\rho)};$$

the general equation of the motion of the atmosphere

will therefore become relative to the laminæ of level, with respect to which δr is very nearly nothing,

$$r^2.\delta\theta.\left\{\left(\frac{d^2u'}{dt^2}\right)-2n.\sin.\theta.\cos.\theta.\left(\frac{dv'}{dt}\right)\right\}$$
$$+r^2.\delta\varpi.\left\{\sin.^2\theta.\left(\frac{d^2v'}{dt^2}\right)+2n.\sin.\theta.\cos.\theta.\left(\frac{du'}{dt}\right)\right.$$
$$\left.+\frac{2n.\sin.^2\theta}{r}.\left(\frac{ds'}{dt}\right)\right\}=\delta V'-g.\delta\varphi-g.\delta y'+n^2r.$$
$$\sin.^2\theta.\delta.(s'-(s')),$$

$\alpha(s')$ being the variation of r corresponding in the state of equilibrium to the variatons $\alpha u'$ and $\alpha v'$ of the angles θ and ϖ.

Let us suppose that all the molecules of air situated at first upon the same terrestrial radius, remain constantly upon the same radius in the state of motion, which has place by what precedes in the oscillations of the sea; and let us try if this supposition will satisfy the equations of the motion and of the continuity of the atmospheric fluid. For this purpose it is necessary, that the values of u' and v' should be the same for all these molecules; but the value $\delta V'$ is very nearly the same for these molecules, as will be seen when we shall determine in the sequel the forces from which this variation results; it is therefore necessary that the variations $\delta\varphi$ and $\delta y'$ should be the same for all these molecules, and moreover that the quantities $2nr.\delta\varpi.\sin.^2\theta.\left(\frac{ds'}{dt}\right)$, and $n^2r.\sin.^2\theta.\delta.\{s'-(s')\}$ may be neglected in the preceding equation.

At the surface of the sea we have $\varphi=y$, αy being the elevation of the surface of the sea above its surface of level. Let us examine if the suppositions of φ equal to y, and of y constant for all the molecules of air situated upon the same radius, can subsist with the equation of the continuity of the fluid. This equation, by No. 35, is

$$0 = r^2 \cdot \left\{ \varrho' + (\rho) \cdot \left\{ \left(\frac{du'}{d\theta}\right) + \left(\frac{dv'}{d\varpi}\right) + \frac{u'.\cos.\theta}{\sin.\theta} \right\} \right\} + (\rho)\left(\frac{d\ r^2 s'}{dr}\right);$$

from which we may obtain

$$y' = -l. \left\{ \left(\frac{d.r^2 s'}{r^2 dr}\right) + \left(\frac{du'}{d\theta}\right) + \left(\frac{dv'}{d\varpi}\right) + \frac{u'.\cos.\theta}{\sin.\theta} \right\}.$$

$r + \alpha s'$ is equal to the value of r of the surface of level which corresponds to the angles $\theta + \alpha u$ and $\varpi + \alpha v$, plus the elevation of the molecule of air above this surface; the part of $\alpha s'$ which depends upon the variation of the angles θ and ϖ being of the order $\dfrac{\alpha n^2 . u}{g}$, may be neglected in the preceding expression of y', and consequently it may be supposed in this expression that $s' = \varphi$; lastly if we make $\varphi = \dot{y}$, we shall have $\left(\dfrac{d\varphi}{dr}\right) = 0$, because the value of φ is then the same relative to all the molecules situated upon the same radius. Moreover y is by what precedes of the order l or $\dfrac{n^2}{g}$; the expression of y will thus become

$$y' = -l. \left\{ \left(\frac{du'}{d\theta}\right) + \left(\frac{dv'}{d\varpi}\right) + \frac{u'.\cos.\theta}{\sin.\theta.} \right\};$$

therefore u' and v' being the same for all the molecules situated originally upon the same radius, the value of y' will be the same for all these molecules. Moreover it is evident from what has been said, that the quantities $2nr.\delta\varpi.\sin.^2\theta \cdot \left(\dfrac{ds'}{dt}\right)$ and $n^2 r.\sin.^2\theta.\delta.\{s' - (s')\}$, may be neglected in the preceding equation of the motion of the atmosphere, which can then be satisfied by supposing that u' and v' are the same for all the molecules of air situated originally upon the same radius: the supposition that all these molecules remain constantly upon the same radius during the oscillations of the fluid, is therefore admissable with the equations of the motion and of the continuity of the atmospheric fluid. In this case the oscillations of divers laminæ of level are the same, and may be determined by means of the equations

$$r^2 . \delta\theta . \left\{ \left(\frac{d^2 u'}{dt^2}\right) - 2n . \sin.\theta . \cos.\theta . \left(\frac{dv'}{dt}\right) \right\}$$

$$+ r^2 . \delta\varpi \left\{ \sin.^2\theta . \left(\frac{d^2 v'}{dt^2}\right) + 2n . \sin.\theta . \cos.\theta . \right.$$

$$\left. \left(\frac{du'}{dt}\right) . \right\} = \delta V' - g . \delta y' - g . \delta y ;$$

$$y' = -l . \left\{ \left(\frac{du'}{d\theta}\right) + \left(\frac{dv'}{d\varpi}\right) + \frac{u' . \cos.\theta}{\sin.\theta} \right\} .$$

These oscillations of the atmosphere ought to produce analogous oscillations in the altitudes of the barometer. To determine these by means of the first, let us suppose a barometer fixed at any height whatever above the surface of the sea. The altitude of the mercury is proportional to the pressure which its surface exposed to that of the air experiences; it may therefore be represented by $lg.\rho$; but this surface is successively exposed to the action of different laminæ of level which elevate and lower themselves like the surface of the sea; thus the value of ρ at the surface of the mercury varies; first, because it appertains to a lamina of level which in the state of equilibrium was less elevated by the quantity αy; secondly, because the density of a lamina is augmented in the state of motion by $\alpha \rho'$ or by $\frac{\alpha(\rho).y}{l}$.

In consequence of the first cause the variation of ρ is

$-\alpha y \cdot \left(\dfrac{d\rho}{dr}\right)$, or $\dfrac{\alpha(\rho) \cdot y'}{l}$; the total variation of the density ρ at the surface of the mercury is therefore $\alpha(\rho) \cdot \dfrac{y+y'}{l}$. It follows from the above, that if the altitude of the mercury in the barometer at the state of equilibrium is denoted by k; its oscillations in the state of motion will be expressed by the function $\dfrac{\alpha k \cdot (y+y')}{l}$; they are therefore similar at all heights above the surface of the sea, and proportional to the altitudes of the barometer.

It now only remains in order to determine the oscillations of the sea and of the atmosphere, to know the forces which act upon these two fluid masses and to integrate the preceding differential equations; which will be done in the fourth book of this work.

CHAP. IX.*

Of the law of universal gravity obtained from phenomena.

38. AFTER having developed the laws of motion, we will proceed to derive from them and from those of the celestial motions presented in detail in the work entitled, Exposition du Systeme du Monde, the general law of these motions. Of all the phenomena that which seems to be the most proper to discover this law, is the elliptic motion of the planets and of comets about the sun: let us see what may be derived from it. For this purpose, let x and y represent the rectangular co-ordinates of a planet in the plane of its orbit, having their origin at the centre of the sun; also let P and Q denote the forces

---

* This chapter which forms part of the first chapter of the second book of the Mechanique Celeste, is added in order to afford the reader some idea of the manner in which Laplace applies the rules given in the introductory treatise.

which act upon this planet parallel to the axes of x and y, during its relative motion about the sun, these forces being supposed to tend towards the origin of the co-ordinates, lastly, let dt represent the element of the time which we will regard as constant; we shall have by Chap. 2,

$$0 = \frac{d^2x}{dt^2} + P; \qquad (1)$$

$$0 = \frac{d^2y}{dt^2} + Q. \qquad (2)$$

If we add the first of these equations multiplied by $-y$ to the second multiplied by x, the following equation will be obtained,

$$0 = \frac{d.(xdy - ydx)}{dt^2} + xQ - yP.$$

It is evident that $xdy - ydx$ is equal to twice the area which the radius vector of the planet describes about the sun during the instant dt; by the first law of Kepler this area is proportional to the element of the time, we shall therefore have

$$xdy - ydx = cdt,$$

c being a constant quantity; the differential of the first member of this equation is equal to nothing, consequently

$$xQ - yP = 0.$$

It follows from this equation that P has to Q the same ratio as x has to y, and that their resultant passes by the origin of the co-ordinates; that is by the centre of the sun. This is otherways evident, for the curve described by the planet is concave towards the sun, consequently the force which causes it to be described tends towards that star.

The law of the proportionality of the areas to the times employed to describe them, therefore conducts us to this remarkable result; that the force which solicits the planets and the comets is directed towards the centre of the sun.

39. Let us now determine the law by which the force acts at different distances from this star. It is evident that the planets and the comets alternately approach to and recede from the sun, during each revolution; the nature of the elliptic motion ought to conduct us to this law. For which purpose resuming the differential equations (1) and (2) of the preceding No., if we add the first multiplied by dx to the second multiplied by dy, we shall have

$$0 = \frac{dx.d^2x + dy.d^2y}{dt^2} + Pdx + Qdy;$$

which gives by integration

$$0 = \frac{dx^2 + dy^2}{dt^2} + 2\int(Pdx + Qdy);$$

the constant quantity being indicated by the sign of integration. If we substitute instead of dt its value $\frac{xdy - ydx}{c}$, which is given by the law of the proportionality of the areas to the times, we shall have

$$0 = \frac{c^2(dx^2 + dy^2)}{(xdy - ydx)^2} + 2.\int(Pdx + Qdy).$$

For greater simplicity let the co-ordinates x and y be transformed into a radius vector and a traversed angle conformably to astronomical practice. Let r represent the radius drawn from the centre of the sun to that of the planet, or its radius vector, and v the angle which it forms with the axis of x; we shall then have

$$x = r.\cos.v; \quad y = r.\sin.v; \quad r = \sqrt{x^2 + y^2};$$

consequently

$$dx^2 + dy^2 = r^2 dv^2 + dr^2; \quad xdy - ydx = r^2 dv.$$

If the principal force that acts upon the planet be denoted by φ; the preceding No. will give

$$P = \varphi.\cos.v; \quad Q = \varphi.\sin.v; \quad \varphi = \sqrt{P^2 + Q^2},$$

therefore

$$Pdx + Qdy = \varphi dr;$$

by substitution we shall have

$$0 = \frac{c^2(r^2 dv^2 + dr^2)}{r^4 dv^2} + 2\int \varphi dr;$$

consequently

$$dv = \frac{cdr}{r.\sqrt{-c^2 - 2r^2 \int \varphi dr}}. \quad (3)$$

This equation will give by means of quadratures the value of v in r, when the force φ is a known function of r; but this force being unknown, if the nature of the curve which it causes to be described is given, by differentiating the preceding expression of $2\int\varphi dr$, we shall have to determine φ the equation

$$\varphi = \frac{c^2}{r^3} - \frac{c^2}{2}.d.\frac{\left\{\frac{dr^2}{r^4 dv^2}\right\}}{dr}. \quad (4)$$

The orbits of the planets are ellipses, having the centre of the sun at one of the foci; if in the ellipse ϖ denotes the angle which the major axis makes with the axis of x, and the origin of x be fixed at the focus, and a represent the semi-major axis, and e the ratio of the eccentricity to the semi-major axis; we shall have

$$r = \frac{a(1 - e^2)}{1 + e.\cos.(v - \varpi)},$$

which equation belongs to a parabola if $e = 1$, and a be infinite; and to an hyperbola if e surpass unity, and a be negative. This equation gives

$$\frac{dr^2}{r^4 dv^2} = \frac{2}{a.(1 - e^2)} \cdot \frac{1}{r} - \frac{1}{a^2(1-e^2)};$$

consequently

$$\varphi = \frac{c^2}{a(1-e^2)} \cdot \frac{1}{r^2};$$

therefore, the orbits of the planets being conic sections, the force φ is inversely as the square of the distance of the centre of these planets from that of the sun.

We also perceive that if the force φ is inversely as the square of the distance, or expressed by $\frac{h}{r^2}$, h being a constant coefficient, the preceding equation of conic sections will satisfy the differential equation (4) between r and v, which gives the expression of φ when we change φ into $\frac{h}{r^2}$. We shall then have $h = \frac{c^2}{a(1-e^2)}$ which forms a conditional equation between the two constant quantities a and e of the equation of conic sections; the three constant quantities a, e, and ϖ of this equation are therefore reduced to two distinct constant quantities, and as the differential equation between r and v is only of the second order, the finite equation of conic sections is the the complete integral

From the above it follows, that if the curve described be a conic section the force is in the inverse ratio of the square of the distance, and reciprocally if the force be inversely as the square of the distance, the curve described is a conic section.

40. The intensity of the force φ relative to each planet and to each comet depends upon the coefficient $\frac{c^2}{a(1-e^2)}$; the laws of Kepler likewise give the means of determining it. Thus, if T denote the time of the revolution of a planet, the area that its radius vector describes during this time being the surface of the planetary ellipse is equal to $\pi a^2 \sqrt{1-e^2}$, π being the ratio

of the semi-circumference to the radius; but by what precedes, the area described during the instant dt is $\frac{1}{2} cdt$: the law of the proportionality of the areas to the times will therefore give the following proportion,

$$\tfrac{1}{2} cdt : \pi a^2 \cdot \sqrt{1-e^2} :: dt : T ;$$

therefore

$$c = \frac{2\pi a^2 \sqrt{1-e^2}}{T}.$$

Relative to the planets, the law of Kepler, that the squares of the times of their revolutions are as the cubes of the great axes of their ellipses, gives $T^2 = k^2 a^3$, k being the same for all the planets; we therefore have

$$c = \frac{2\pi \sqrt{a(1-e^2)}}{k}.$$

$2a(1-e^2)$ is the parameter of the orbit, and in different orbits the values of c are as the areas traced by the radii vectores in equal times; these areas are therefore as the square root of the parameters of the orbits.

This proportion has equally place relative to the orbits of comets compared either to each other or to the orbits of the planets; this is one of the fundamental points of their theory which answers so exactly to all their observed motions. The major axes of their orbits and the times of their revolutions being unknown, we calculate the motions of these stars in a parabolic orbit, and, expressing by D their perihelion distance, we suppose $c = \frac{2\pi \sqrt{2D}}{k}$; which is equivalent to making e equal to unity, and a infinite, in the preceding expression of c; we have therefore relative to comets, $T^2 = k^2 a^3$, so that when their revolutions shall be known, the major axes of their orbits can be determined.

The expression of c gives

$$\frac{c^2}{a(1-e^2)} = \frac{4\pi^2}{k^2},$$

we have therefore

$$\varphi = \frac{4\pi^2}{k^2} \cdot \frac{1}{r^2}.$$

The coefficient $\frac{4\pi^2}{k^2}$ being the same for all the planets and comets, it results that for each of these bodies, the force φ is inversely as the square of the distances from the centre of the sun, and only varies from one body to another by reason of these distances; from which it follows, that it is the same for all these bodies supposed at equal distances from the sun.

We are therefore conducted by the beautiful laws of Kepler to regard the centre of the sun as the focus of an attractive force which extends itself infinitely in all directions, decreasing in the ratio of the squares of the distances. The law of the proportionality of the areas described by the radii vectores to the times employed in describing them, proves to us that the principal force which solicits the planets and the comets is constantly directed towards the centre of the sun; the ellipticity of the planetary orbits, and the very nearly parabolical motions of the comets, shew that for each planet and for each comet this force is inversely as the square of the distance of these stars from the sun; lastly, from the law of the proportionality of the squares of the times of the revolutions to the cubes of the major axes of the orbits, or that of the proportionality of the areas described during the same time by the radii vectores in different orbits to the square roots of the parameters of these orbits, which law contains the preceding and is extended to comets; it results, that this force is the same for all the planets and the comets placed at equal distances from the sun, so that in this case, these bodies would be precipitated towards it with the same velocity.

ERRATA.

Page 11. Notes line 3, for, *direction by S. &c.* read, *direction by lines denoted by S. &c.*

Page 56. Line 1, for, dx, read, d^2x.

Page 135. Notes line 4 from the bottom, instead of, *to one half that,* read, *to that.*